Aldo-Keto Reductases
and Toxicant Metabolism

ACS SYMPOSIUM SERIES **865**

Aldo-Keto Reductases and Toxicant Metabolism

Trevor M. Penning, EDITOR
University of Pennsylvania School of Medicine

J. Mark Petrash, EDITOR
Washington University School of Medicine

**Sponsored by the
ACS Division of Chemical Toxicology**

American Chemical Society, Washington, DC

Library of Congress Cataloging-in-Publication Data

Aldo-keto reductases and toxicant metabolism / Trevor M. Penning, J. Mark Petrash, editors.

   p. cm.—(ACS symposium series ; 865)

  "Sponsored by the ACS Division of Chemical Tolicology. "

  Includes bibliographical references and index.

  ISBN 0–8412–3846–4

  1. Metabolic detoxification—Congresses. 2. Carbonyl reductase—Congresses.

 I. Penning, Trevor M., 1951- II. Petrash, J. Mark, 1955-. III. Series.

RA1220.A43   2003
612′.015191—dc22   2003056090

The paper used in this publication meets the minimum requirements of American National Standard for Information Sciences—Permanence of Paper for Printed Library Materials, ANSI Z39.48–1984.

PRINTED IN THE UNITED STATES OF AMERICA

Foreword

The ACS Symposium Series was first published in 1974 to provide a mechanism for publishing symposia quickly in book form. The purpose of the series is to publish timely, comprehensive books developed from ACS sponsored symposia based on current scientific research. Occasionally, books are developed from symposia sponsored by other organizations when the topic is of keen interest to the chemistry audience.

Before agreeing to publish a book, the proposed table of contents is reviewed for appropriate and comprehensive coverage and for interest to the audience. Some papers may be excluded to better focus the book; others may be added to provide comprehensiveness. When appropriate, overview or introductory chapters are added. Drafts of chapters are peer-reviewed prior to final acceptance or rejection, and manuscripts are prepared in camera-ready format.

As a rule, only original research papers and original review papers are included in the volumes. Verbatim reproductions of previously published papers are not accepted.

ACS Books Department

Contents

vii

Aldo-Keto Reductases and Exogenous Toxicants: Mycotoxins, Aldehydes, and Ketones

Aldo-Keto Reductases, the Stress Response, and Cell Signaling

Indexes

Preface

The past decade has seen rapid growth in research on aldo-keto reductases (AKR), a superfamily of NADPH-dependent oxidoreductases that share structural and functional features with one another, and are ubiquitous in a wide range of organisms, including bacteria, yeasts, and man. Human tissues contain many different forms of aldo-keto reductases, but their intrinsic physiological functions are still poorly understood. However, AKRs have long been associated with the polyol pathway in which high circulating glucose is converted to sorbitol. The accumulation of this hyperosmotic sugar may contribute to the complications arising from diabetes, and has led the search for aldose-reductase inhibitors. Growing evidence exists that AKR enzymes play a role in the detoxication of aldehydes and ketones, including the deactivation of the important reactive aldehydes derived from the endogenous oxidation of lipids. Further, AKR enzymes are of great interest because they contribute to the metabolism of prostaglandins, drugs, steroid hormones, and environmental chemical carcinogens, including aflatoxins and carcinogens in tobacco smoke. In addition, finally, evidence exists that AKRs function as modulators of stress response associated with the exposure of cells to toxins and drugs.

Research on AKR enzymes is very significant because aldehydes and ketones are ubiquitous in cellular environments because they are by-products of various endogenous metabolic oxidative processes and because many environmental contaminants are precursors of these reactive intermediatives in biological systems. The keto group is chemically reactive and therefore has the potential of reacting with biological macromolecules, thus resulting in posttranslational modifications of proteins, or the formation of DNA adducts. Depending on their chemical stabilities and structures, and if not removed by DNA repair enzymes, DNA lesions can cause mutations upon DNA replication. If these mutations occur at critical sites in tumor suppressor or growth-activating genes, these DNA lesions can contribute to the initiation stages of cancers.

The potential contributions of aldo-keto reductases in the etiology of cancers associated with exposure to tobacco specific nitrosamines, mycotoxins, and polycyclic aromatic hydrocarbons (PAH) has recently elicited considerable interest in the chemical carcinogenesis research community. For example, the metabolic activation of PAH by cytochrome P450 enzymes (CYP) to *trans*-dihydrodiols, and their further transformation to highly genotoxic diol epoxides, has long been considered as the most important metabolic pathway of activation of this class of chemical carcinogens. However, now a growing appreciation exists that the AKR enzymes may also contribute significantly to the oxidation of the *trans*-dihydrodiol PAH generated by CYPs to yield *ortho-quinones*. The latter are highly reactive and readily form not only potentially mutagenic DNA adducts, but also contribute to cellular oxidative stress by generating reactive oxygen species via redox cycling pathways.

Our understanding of the structural and functional properties of AKRs has grown immensely within the past few years, and research in this area is still accelerating at a great pace. This emerging trend in toxicology research was noted with great interest by the Division of Chemical Toxicology, and it was concluded that a full symposium at a National American Chemical Society (ACS) meeting would be most timely and elicit considerable interest. Trevor Penning, a leader in the field of aldo-keto reductases, was invited to organize a symposium on this topic at the 224th National Meeting of the ACS that was held on August 20, 2002, in Boston, Massachusetts. Given the tremendous response and interest in this burgeoning area of AKR research, the ACS has selected this Symposium for publication in the ACS Symposium Series. The ACS Division of Chemical Toxicology is proud to sponsor this book, the first one to be published under the aegis of this still young ACS Division.

The 16 chapters in this book, all written by well-recognized leaders in their respective research areas, provide authoritative overviews of this emerging field. Knowledgeable researchers, new investigators, students, and the interested casual reader will find, all in one place, not only authoritative and informative accounts of the status of each subject area, but also insights into the significance of the latest research findings. This book will fill an important void because of a lack of recently published, comprehensive, and wide-ranging reviews on aldo-keto reductases. There is no question that the publication of this volume will be viewed as an important milestone in the field. I fully expect that this comprehensive book will serve as an invaluable resource on AKR research for both new and seasoned investigators and their students for many years to come!

The editors of this book are indebted to Gail Vogel, whose editorial skills made this symposium book possible.

Nicholas E. Geacintov

Program Chair, 2000–2002
ACS Division of Chemical Toxicology
Chemistry Department
New York University
New York, NY 10003-5189
Email: ng1@nyu.edu

General Overview

.

Chapter 1

Introduction and Overview of the Aldo-Keto Reductase Superfamily

Trevor M. Penning

Department of Pharmacology, University of Pennsylvania, Philadelphia, PA 19104–6084

Aldo-Keto Reductases (AKRs) are a rapidly growing gene superfamily involved in the formation of hyperosmotic sugars, which can contribute to diabetic complications. They also metabolize reactive aldehydes, prostaglandins, steroid hormones, chemical carcinogens and drugs. These enzymes (>115) are generally monomeric, cytosolic, NAD(P)H-dependent- oxidoreductases, that share high sequence identity (<40%) and similar three-dimensional-structures e.g., $(\alpha/\beta)_8$-barrels and belong to 14 families. For a complete listing of the 14 families visit: www.med.upenn.edu/akr. AKRs detoxify endogenous toxicants, e.g., cytotoxic/mutagenic bifunctional electrophiles derived from the decomposition of lipid hydroperoxides. AKRs also metabolize exogenous toxic substances. They play a role in tobacco-carcinogenesis since they catalyze the detoxication of nicotine derived nitrosamino-ketones and activate polycyclic aromatic *trans*-dihydrodiols to yield reactive and redox active *ortho*-quinones. Discrete isozymes also detoxify mycotoxin metabolites, e.g., aflatoxin dialdehyde and protect against hepatocellular carcinoma. The aflatoxin reductases are induced by cancer chemopreventive agents, e.g., ethoxyquin. Evidence is mounting that AKRs are stress regulated genes and play a central role in the cellular response to osmotic, electrophilic and oxidative stress.

Introduction

Members of the Aldo-Keto Reductase (AKR) gene superfamily are expressed in prokaryotes and eukaroytes and are highly conserved (*1-3*). In mammals they are involved in the metabolism of endogenous toxicants (sugar and lipid derived aldehydes) (*4,5*) and exogenous toxicants (carcinogen metabolites e.g., tobacco specific nitrosamino-ketones (4-(*N*-methyl-*N*-nitrosamino)-1,3-pyridyl)-1-butanone (NNK), PAH *trans*-dihydrodiols and aflatoxin dialdehyde (*6-8*). Many of the reactions are characterized by the reduction of carbonyl groups to alcohols. Because of their broad substrate specificity AKRs may be ultimately as important as the CYP superfamily in drug and xenobiotic metabolism. Yet little is known about their drug substrates. Individual variation in the expression of AKR isoforms may ultimately determine susceptibility to drug, carcinogen and toxicant exposure. The aim of this article is to provide an overview of the AKRs and their role in toxicology and the stress response.

Properties of the Aldo Keto Reductases

AKRs are predominantly monomeric NAD(P)(H) dependent oxidoreductases. They convert carbonyl groups to alcohols so that the products can often undergo conjugation and elimination. When this reaction occurs on drugs AKRs can play an important role in functionalizing these agents for clearance. Their monomer molecular weight = 34-37 kDa. There are currently 115 members in the superfamily that fall into 14 families (where less than 40% sequence identity defines membership to a unique family). In addition to these proteins, there are greater than 100 potential members based on genome projects. A complete list of AKR members can be found at www.med.upenn.edu/akr (*1-3*), Figure 1.

Based on their broad substrate specificity there is debate as to the identity of the natural substrates for some AKRs. Endogenous substrates include but are not limited to: sugar aldehydes, steroid hormones, prostaglandins and lipid derived aldehydes. Exogenous substrates include tobacco-derived carcinogens (tobacco specific nitrosamino-ketones (NNK), and PAH *trans*-dihydrodiols) and metabolites of mycotoxins (aflatoxin dialdehyde).

Over 15 crystal structures of AKRs complexed with their ligands have been determined. Each crystal structure adopts an $(\alpha/\beta)_8$-barrel structure and contain two additional helices. In this motif there is an alternating arrangement of α-helix and β-strand that repeats itself 8-times. The β-strands coalesce at the core of the structure to form the staves of the barrel. At the base of the barrel there is a highly conserved catalytic tetrad (Tyr 55, Asp 50, Lys 84, His 117, numbering based on AKR1C9). The AKR website is linked to the protein database (PDB) which allows complete rendering of the deposited 3-dimensional structures.

Aldo-Keto Reductase Superfamily Members

- Mainly Monomeric NAD(P)(H) Dependent
  Oxidoreductases
  - convert carbonyl ◄——► alcohol
  -M.W. 34-37 kDa (monomers)
  -115 members fall into 14 families

- Substrate Specificity

  -sugar aldehydes
  -steroid hormones
  -prostaglandins & lipid aldehydes
  -chemical carcinogens
      NNK
      PAH *trans*-dihydrodiols
      Aflatoxin dialdehyde

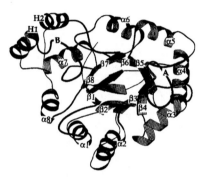

(www.med.upenn.edu/akr)

- Conserved Catalytic Tetrad
  Tyr 55, Asp 50, Lys 84, His 117

- $(\alpha,\beta)_8$-Barrel Structure

Figure 1. Properties of A̲ldo-K̲eto R̲eductase superfamily members

Each AKR catalyzes an ordered bi-bi kinetic mechanism in which cofactor binds first and leaves last (*9-11*). For several members of the superfamily it is the rate of $NADP^+$ release that is often rate-limiting (*10,11*). Loop structures (Loops A, B and C at the back of the barrel) define AKR substrate specificity. A common catalytic mechanism exists in which there is 4-pro-*R* hydride transfer from the pyridine nucleotide cofactor to the acceptor carbonyl facilitated by a general acid/base (*12,13*). The catalytic tyrosine functions as both the general acid and base (*12,13*).

Superfamily Structure and Human Members

Cluster analysis shows that 14 AKR families exist where members of individual families are less than 40% identical to members of other families. The current dendrogram shows early divergence from ancestral proteins which divides the superfamily into two groups: AKR1 – AKR5 is one group while the other group is AKR6-AKR14 (*14*), Figure 2.

Less than 40% amino acid sequence identity assigns members to individual families. While greater than 60% amino acid sequence identity assigns a protein to a subfamily. The nomenclature system uses the <u>AKR</u> root to identify the protein as an aldo-keto reductase; the first numeral assigns the protein to one of the 14 families AKR<u>1</u>; the first letter after the numeral assigns the protein to a unique subfamily AKR1<u>A</u>; and the last numeral assigns the unique protein AKR1A<u>1</u>. The last numeral is assigned in chronological order of discovery. By far the largest family is the AKR1 family which contains, aldehyde reductase (AKR1A1), aldose reductase (AKR1B1), hydroxysteroid dehydrogenases (AKR1C subfamily), and the steroid 5β-reductases (AKR1D subfamily). The AKR6 family contains the β-subunit of the voltage-dependent potassium channel. These subunits do not have enzymatic activity but modulate opening of the potassium channel in the presence of added cofactor. The AKR7 family contains the aflatoxin aldehyde reductases. Crystal structures of the rat aflatoxin aldehyde reductase show that it is a dimeric protein (*15*). Both homo- and heterodimers are possible where (AKR7A1: AKR7A4 (1:1)) would indicate a heterodimer with the stoichiometry shown (*16*). Evidence exists for dimer formation in the AKR2, AKR6 and AKR7 families.

The human genome project (HUGO) has adopted the AKR nomenclature system and currently thirteen human AKRs are known to be functionally expressed, Table 1. Not surprisingly, many of the proteins belong to the AKR1 family and include the human homologs of aldose reductase, aldehyde reductase, hydroxysteroid dehydrogenases, and steroid 5β-reductases. Other members include the potassium channel subunits and the human homologs of aflatoxin aldehyde reductase.

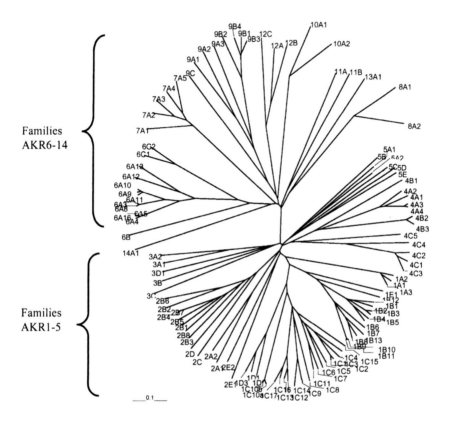

Figure 2. Dendrogram of current AKR family members

Table I. Human AKR Members Identified by HUGO

Gene	Protein	Chromosomal Localization
AKR1A1	Aldehyde reductase	1p33-p32
AKR1B1	Aldose reductase	7q35
AKR1B10	Aldose reductase	
AKR1C1	20α-HSD, dihydrodiol dehydrogenase 1 (DD1)	10p15-10p14
AKR1C2	Type 3, 3α-HSD (DD2)	10p15-10p14
AKR1C3	Type 2, 3α-HSD, Type 5 17β-HSD, (DDX)	10p15-10p14
AKR1C4	Type 1, 3α-HSD (DD4)	
AKR1D1	5β-Reductase	
AKR6A3	Potassium voltage gated channel, β-subunit-1	3q26.1
AKR6A5	Potassium voltage gated channel, β-subunit-1	1p36.3
AKR6A9	Potassium voltage gated channel, β-subunit-1	17p13.1
AKR7A2	Aflatoxin aldehyde reductase	1p35.1—p36.23
AKR7A3	Aflatoxin aldehyde reductase	

The emphasis in this overview is on those proteins that are involved in the metabolism of endogenous and exogenous toxicants as a preview to other articles. The focus will be on aldose and aldehyde reductase and their roles in the metabolism of sugar and lipid derived aldehydes. In addition, the focus will be on the role of AKRs in the metabolism of exogenous toxicants, including tobacco derived carcinogens (nitrosamines and polycyclic aromatic hydrocarbons) and hepatocarcinogens (aflatoxin). What emerges is that AKRs are often upregulated in response to stressors in an attempt to protect the cell from these various insults.

AKRs and Endogenous Toxicants: Detoxication of Aldehydes in Diabetes

Traditionally, human aldose reductase (AKR1B1) was implicated in diversion of glucose (sugar aldehyde) to sorbitol (polyol) pathway (*4*), Figure 3. In the diabetic more glucose is shunted down the polyol pathway to form sorbitol. Unfortunately, sorbitol is a hyperosmotic sugar and its accumulation may be responsible for the complications arising from diabetes (e.g. caratogenesis, retinopathy, neuropathy and nephropathy) (*17*). As a result there has been a push to develop aldose reductase inhibitors (e.g. tolrestat and sorbinil) to prevent the complications arising from diabetes. However, glucose is a relatively poor substrate for AKR1B1 suggesting that this enzyme may have other functions (*18*).

Other reactive aldehydes are also elevated in diabetic patients and are key intermediates in the formation of advanced glycation end products and may also contribute to diabetic complications. These include glyoxal, methyl glyoxal, D-glucosone and 4-hydroxy-2-nonenal. Many of these compounds are superior substrates to glucose suggesting a more important role for AKR1B1 is to provide protection against other toxic reactive aldehydes. This topic is discussed in the article by Dr. Vander Jagt, Figure 3.

AKRs and Endogenous Toxicants: Detoxication of Lipid Aldehydes

Polyunsaturated fatty acids (PUFA's) found in the plasma membrane (e.g. linoleic acid and arachidonic acid) can undergo attack by reactive oxygen species (ROS) to produce lipid hydroperoxides (e.g., 13-hydroperoxy-octadecanoic acid (13-HPODE) and 15-hydroperoxyeicosatetraenoic acid (15-HPETE) (19,20). Decomposition of these lipid hydroperoxides results in the formation of α,β-unsaturated aldehydes (e.g. 4-hydroxy-2-nonenal (4-HNE)) which are reactive bifunctional electrophiles that can react with nucleophiles on macromolecules e.g. protein and DNA.

In the case of 4-HNE, the aldehyde can form both a Schiff's base with the ε-amino group of lysine and undergo 1,4-Michael addition with cysteines, hence it is a bi-functional electrophile. The formation of 4-HNE immunoreactive adducts have been observed in artherosclerotic plaques (21), in degenerating neurons in the substania nigra in Parkinson's disease (22) and in neurofibrillary tangles in Alzheimer's disease (23). AKRs that can reduce 4-HNE to the nontoxic 1,2-dihydroxynonene would play an important role in the detoxication of this reactive aldehyde, Figure 4.

Several AKRs have been implicated in this process including AKR1A1 (aldehyde reductase), AKR1B1 (aldose reductase) and AKR1C1-AKR1C3 (5,18). Of these AKR1C1 is inducible by ROS which would aid in the elimination of 4-HNE (24). In addition, AKR1B1 has a high catalytic efficiency for the glutathionyl conjugates of reactive aldehydes and is involved in the metabolism of 4-HNE-GS conjugates in isolated perfused heart (25,26). Thus, several AKRs play a role in the elimination of 4-HNE irrespective of GSH status. Collectively, AKRs will contribute to the elimination of GS-conjugates of 4-HNE and provide a cytosolic barrier to 4-HNE when GSH levels have been depleted. The preference of aldose reductase (AKR1B1) to reduce GS-conjugates of 4-HNE is discussed in the article by Dr. Satish Srivastava. In the same article, post-translational modification of a redox-sensitive Cys298 is implicated in altering the catalytic activity of AKR1B1 and its sensitivity to aldose-reductase inhibitors. In a subsequent article by Dr. Sanjay Srivastava, it is shown that AKR1B1 will play a significant role in the metabolism of 4-HNE in human aortic endothelial cells and that it will detoxify oxidized phospholipids

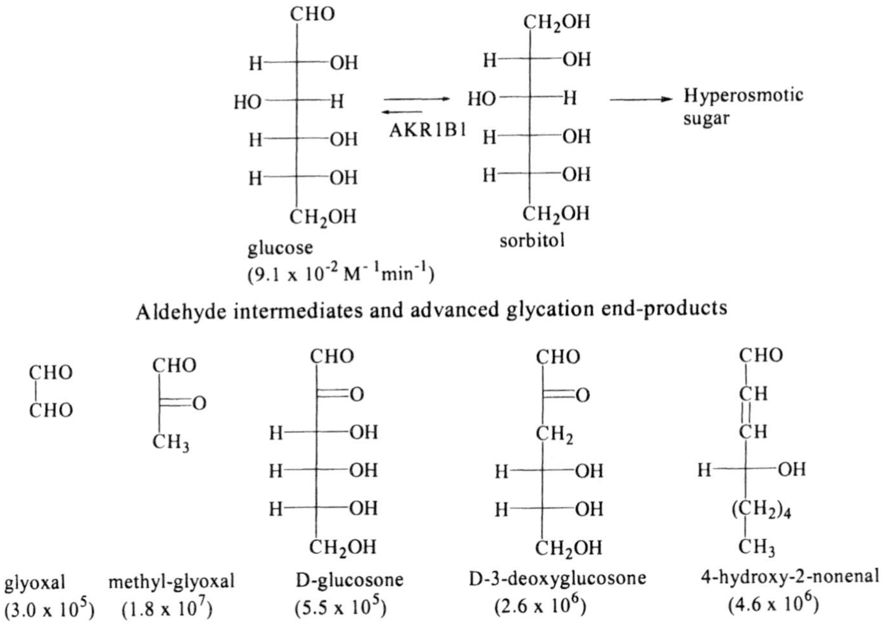

Aldehyde intermediates and advanced glycation end-products

Figure 3. Role of Aldose Reductase (AKR1B1) in the detoxication of aldehydes in diabetes where kcat/Km (M^{-1} min^{-1}) is shown in parenthesis

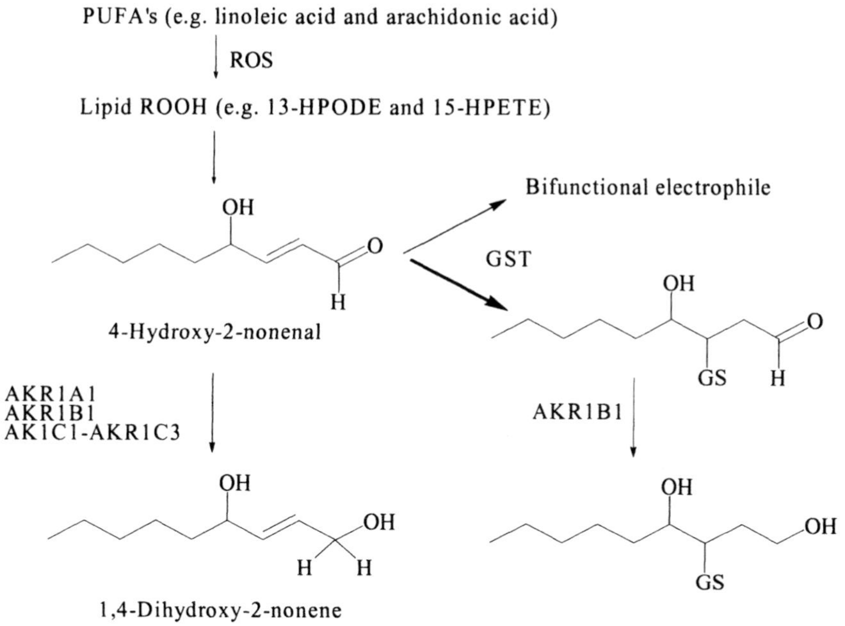

PUFA's (e.g. linoleic acid and arachidonic acid)

| ROS

Lipid ROOH (e.g. 13-HPODE and 15-HPETE)

Figure 4. Role of human AKRs in the detoxication of lipid aldehydes

e.g. 1-palmitoyl-2-(5-oxovaleroyl) phosphatidyl choline. These data indicate that AKR1B1 not only protects against lipid peroxidation but may have atherosclerotic protective potential, as well.

AKRs and Exogenous Toxicants: Tobacco Related Carcinogens

AKRs not only protect from exogenous toxicants (glucose, reactive lipid aldehydes, etc.) but they also are involved in the metabolism of exogenous toxicants. For example, two major carcinogens present in tobacco are nicotine derived nitrosamines and the polycyclic aromatic hydrocarbons (PAH) (*27*). AKRs are involved in the metabolism of both and may impact tobacco carcinogenesis or human lung cancer.

AKRs and NNK Metabolism

NNK (a tobacco specific nitrosaminoketone) undergoes α-hydroxylation catalyzed by CYP2A6/CYP3A4 to form reactive electrophiles (*28*), Figure 5.

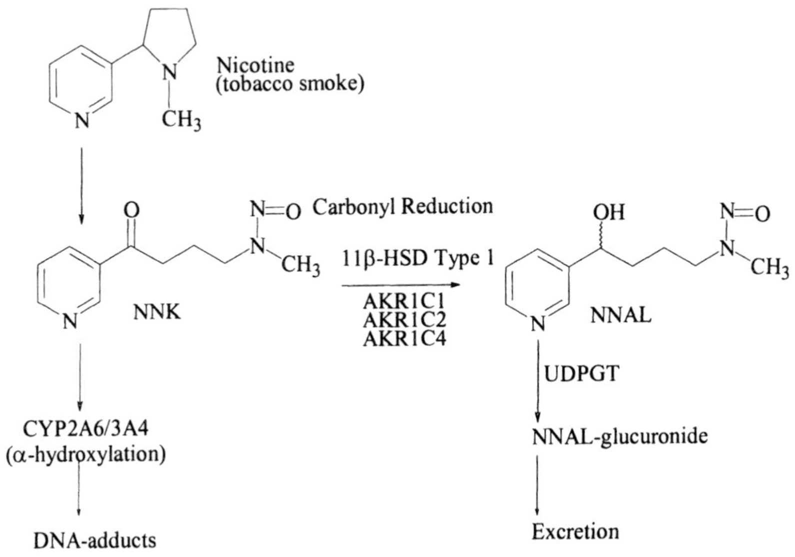

Figure 5. Role of human AKRs in the metabolism of NNK

α-Methylene hydroxylation results in the formation of a methyl diazohydroxide that decomposes to form a methyl carbonium ion to yield methylated DNA adducts (*28*). Alternatively, α-methyl hydroxylation ultimately yields pyridyloxobutyl DNA-adducts (*28*). Carbonyl reduction of NNK to yield the corresponding nitrosoalcohol 4-(*N*-methyl-*N*-nitrosamino)-1-(3-pyridyl)-1-butanol (NNAL) represents the first step in detoxication. Although, NNAL can be subjected to the same α-hydroxylation events, it is also a substrate for glucuronidation, and the NNAL-glucuronide can then be excreted (*6*). Enzymes involved in carbonyl reduction of NNK include 11β-HSD type 1 (a short chain dehydrogenase/reductase), as well as AKR1C1-AKR1C4. The relative contributions of these enzymes to detoxication of NNK will be governed by the balance of expression of these enzymes versus the expression of CYP isoforms and the relevant UDPG-transferases (*6*). This topic is discussed in the article by Drs. Maser and Breyer-Pfaff.

AKRs and PAH Metabolism/Activation

Tobacco smoke also contains a complex mixture of polycyclic aromatic hydrocarbons (*27*). One of the most abundant complete carcinogens present is benzo[*a*]pyrene (BP). Three pathways for the metabolic activation of PAH have been proposed, Figure 6.

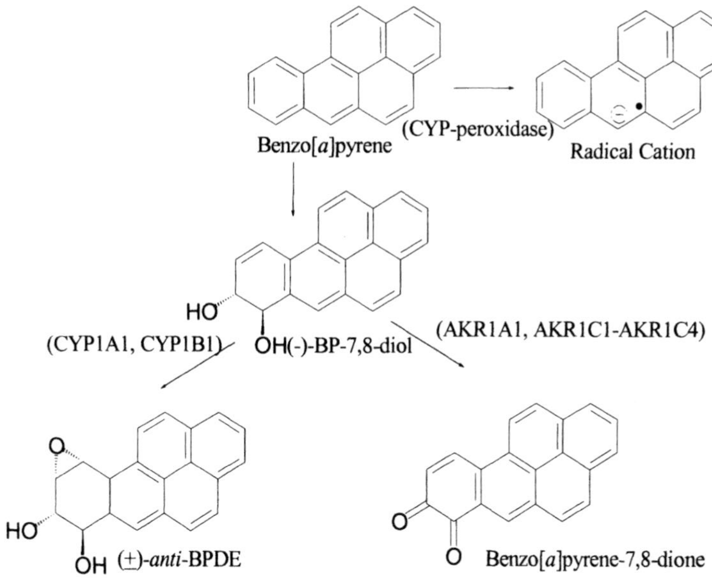

Figure 6. Role of human AKRs in PAH activation

The first pathway involves the CYP peroxidase reaction; during peroxide bond cleavage higher oxidation states of iron are generated which abstract a hydrogen from the most electron rich carbon of BP to yield a radical cation at C6. The radical cation is highly electrophilic and will form depurinating DNA adducts an initiating event (*29*). In the second pathway, the sequential actions of CYP1A1 and epoxide hydrolase form a *trans*-dihydrodiol intermediate, (-)-BP-7,8-diol (a proximate carcinogen). Subsequent monoxygenation yields a highly reactive diol-epoxide, (+)-*anti*-BPDE (*30,31*). Based on its mutagenicity and tumorigenicity (+)-*anti*-BPDE is an ultimate carcinogenic species (*32,33*). In the third pathway of activation, AKRs divert PAH *trans*-dihydrodiols to yield highly reactive and redox-active PAH *o*-quinones, e.g. BP-7,8-dione (*34*). Its electrophilic and redox properties indicates that these metabolites may act as initiators and promoters and may explain the complete carcinogenicity of PAH. AKRs implicated in this pathway in human are aldehyde reductase AKR1A1, and AKR1C1-AKR1C4 (*35,36*). Apart from AKR1C4 all enzymes are expressed in lung tissue, but importantly AKR1C1 is induced by PAH and ROS suggesting that the balance of expression may favor this isoform upon prolonged PAH exposure (*24*). In this example, AKRs in an attempt to eliminate proximate carcinogens form more reactive species. This pathway of metabolic activation is described in the article by Dr. Penning and is followed by articles: on the cloning and expression of murine dihydrodiol dehydrogenases (Dr. Deyashiki); the synthesis of substrates and products of AKR mediated PAH metabolism (Dr.

Harvey); the chemistry of PAH *o*-quinones (Dr. Gopishetty) and oxidative lesions derived from the AKR pathway of PAH activation (Dr. Blair).

AKRs and Exogenous Toxicants: Mycotoxins

Aflatoxin B_1 is a mycotoxin and a potent rodent and human hepato-carcinogen. Aflatoxin is metabolically activated by CYP3A4 to yield the 8,9-epoxide, this electrophilic species can form DNA-adducts (initiation event) GSH-adducts (elimination) or it can hydrolyze to yield a dihydrodiol (*37* and refs. therein), Figure 7.

Rearrangement of the aflatoxin B_1 dihydrodiol yields the highly reactive aflatoxin B_1-dialdehyde which is capable of forming lysine adducts with proteins. Treatment of rats with the anti-oxidant chemopreventive agent ethoxyquin led to the induction of rat liver aflatoxin aldehyde reductase (AKR7A1) and the prevention of aflatoxin induced hepatocarcinogenesis (*8*). Prevention of hepatocarcinogenesis is likely achieved by the induction of glutathione-S-transferase and AKR7A1. AKR7A1 reduces aflatoxin B_1 dialdehyde to a mixture of monoaldehydes which are converted to dialcohols and thus lysine adduct formation is prevented. Subsequently, a non-inducible isoform was discovered in rat AKR7A4 (*38*). AKR7A1 was found to be dimeric and is capable of forming homodimers with itself or heterodimers with AKR7A4 (*16*). In humans, the relevant homologs are AKR7A2 and AKR7A3 *(39,40)*. The roles of rat and human AKR7A family members in the metabolism of aflatoxin B_1-dialdehyde are discussed in articles by Dr. John Hayes and the kinetics and chemistry of lysine adduct formation with the dialdehyde are discussed by Dr. Guengerich.

AKRs: Primordial Genes Regulated by Primordial Signals

AKRs are expressed in prokaryotes through to eukaryotes and show a high conservation in amino acid sequence and three-dimensional topology. An underlying theme is that the enzymes reduce carbonyls, often on toxic aldehydes suggesting that they play important protective roles. This may also explain the redundancy that exists in their overlapping substrate specificity. As part of this protective response they may also inadvertently become engaged in the metabolic activation of carcinogens, as is the case with polycyclic aromatic hydrocarbons.

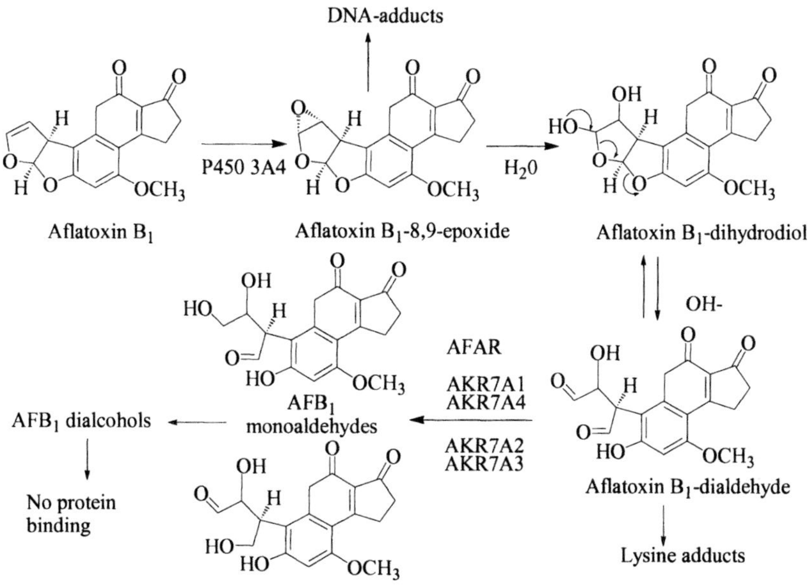

Figure 7. Role of AKRs in the metabolism of aflatoxin

Because these enzymes assume a protective function it is not surprising that they are up-regulated by stress signals e.g., osmotic shock, ROS and electrophiles. For example, AKR1B1 produces hypertonic/hyperosmotic sugars using glucose as a substrate. The AKR1B1 gene is upregulated by tonicity e.g. NaCl where increased AKR1B1 activity would cause tissues to maintain osmolarity by converting glucose to sorbitol (*41,42*), Figure 8. The regulatory element on the gene promoter is referred to as the Osmotic Response Element (ORE) and the transcription factors that bind to the ORE are likely to be regulated by a MAP Kinase module (*43*). Novel transcription factors implicated indirectly in the ORE response may be NF-1 or TONEBP (*44-46*).

Other examples of stress induced AKRs include AKR1B1 (aldose reductase) and AKR1C1. AKR1B1 is induced under myocardial ischemia/reperfusion injury a condition that generates ROS and 4-HNE (*47*). Both NO and PKC signaling pathways have been implicated in the up-regulation of AKR1B1 (*47*). PKC activation ultimately leads to increased AP-1 activity (*48,49*). AKR1C1 implicated in the detoxication of NNK and the metabolic activation of PAH is also up-regulated by reactive oxygen including hydrogen peroxide (*24*), where the AKR1C1 promoter has a cassette of six AP-1 sites (Lin and Penning unpublished observations).

16

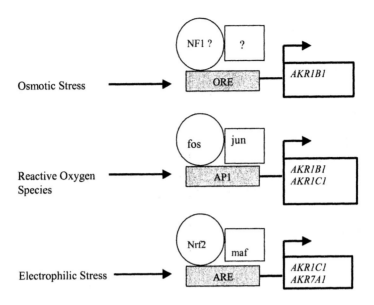

Figure 8. AKRs: primordial genes regulated by primordial signals

Several AKRs involved in carcinogen metabolism (AKR1C1 and AKR7A1) are induced by monofunctional enzyme inducers (anti-oxidants, and electrophiles) by a mechanism consistent with gene regulation via a EpRE/ARE (electrophilic response element/anti-oxidant response element) *(8,24,50)*. Although the functional identity of the EpRE/ARE on the AKR1C1 and AKR7A1 promoters remains to be demonstrated it is anticipated that these genes will be regulated by heterodimers of Nrf2 and maf binding to an EpRE/ARE *(51)*, see Figure 8. Because AKRs exist in prokaryotes and eukaryotes and are regulated by common stress signals, osmotic, electrophilic, and ROS it is speculated that they are primordial genes regulated by primordial signals.

AKRs, the Stress Response and Cell Signaling

Does the regulation of AKRs by stress signals translate into cytoprotective effects? In an *in vivo* myocardial ischemia/reperfusion model in rabbits, in which there are brief periods of ischemia (ischemic preconditioning) there is elevation of AKR1B1 expression (47). Inhibition of AKR1B1 with tolrestat or sorbinil abrogated the infarct-sparing effects of the preconditioning, suggesting that *in vivo* AKR1B1 has a cardioprotective function. AKR1B1 inhibition resulted in the accumulation of 4-HNE suggesting that the cardioprotection is achieved by the AKR1B1 dependent elimination of 4-HNE. This is one of the first examples of how AKRs mediate the response to stress *in vivo*. This topic is discussed in the article by Dr. Bhatnagar.

The ability of AKRs to be regulated by stressors and eliminate stress signals suggests that they may affect signal transduction cascades. In the article by Dr.Romana, inhibition of aldose-reductase (AKR1B1) blocks growth factor (TNFα) signaling pathways upstream from PKC, which in turn blocks the activation of redox-sensitive transcription factors NF-kB and AP-1. *In vivo*, these stimuli are key regulators of vascular smooth muscle cell proliferation and apoptosis of vascular endothelial cells. Thus tolrestat and sorbinil attenuate TNFα stimulated vascular smooth muscle cell proliferation.

Although evidence is mounting that AKRs are stress regulated genes, eliminate stress signals and may affect stress-regulated signal transduction pathways, it has been difficult to determine their physiological significance due to the redundancy that exists in the substrate specificity of AKRs. Yeast provides a powerful model in which to examine the functional significance of AKRs in response to stress. Isogenic strains of yeast have been developed containing multiple AKR gene deletions. A triple AKR null strain was found to have enhanced sensitivity to heat shock. The utility of this approach in delineating the roles of AKRs in responding to stress signals is discussed in the article by Dr. Petrash.

The response of AKRs to stress signals may be a double-edged sword. For example, elevation of AKR1B1 and AKR1C1 by ROS would help effectively eliminate the cytotoxic aldehyde 4-HNE. However, in response to exogenous toxicants the outcome may be different. For example, on exposure to PAH, AKR1C1 makes reactive and redox-active PAH *o*-quinones. Both PAH *o*-quinones and ROS induce AKR1C1. Thus, in response to PAH a positive feedback loop is generated so that more AKR1C1 leads to an elevation of electrophilic and ROS stress leading to cytotoxicity and mutagenicity.

Future Directions

AKRs catalyze the interconversion of carbonyl groups with alcohols on a wide range of substrates. Broad overlapping substrate specificity exists for many substrates. This built in redundancy serves the purpose of ensuring that there are multiple enzymes available to deal with toxicant insult. More effort now needs to be exerted in assigning specific reactions to AKR members and relating this to their tissue specific distribution and whether the isoform is constititively expressed or induced. Because of the widespread presence of carbonyl groups on drugs and the fact that many drugs are metabolized to alcohols for conjugation and elimination it is remarkable that little is known about the drug substrates for each of the human AKRs. AKRs and GST's are similar in that induction of these enzymes by electrophiles may offer chemoprevention against carcinogens (e.g., aflatoxin) but the same induction mechanisms may also establish chemotherapeutic drug resistance. AKR1C1 was identified in ethacrynic acid resistant colon carcinoma cells (*52*). Similarly, induction of AKR1A1 (also danorubicin reductase) could account for resistance to this drug (*53*).

The focus of this overview has been to highlight properties of the AKRs as they pertain to detoxication/toxication mechanisms. However, it should be emphasized that discrete isofroms may be involved in regulating ligand access to nuclear receptors. AKRs that have hydroxysteroid dehydrogenase activity can either form or eliminate active steroid hormone in target tissues and therefore may regulate the occupancy and *trans*-activation of steroid receptors (*54*). A similar case could be made for AKRs that have prostaglandin reductase activity since they may regulate ligand access to PPARγ receptors (*55*). This area needs further attention.

The signal transduction cascades responsible for regulating AKR gene expression in response to stress need to be thoroughly identified. In addition the pharmcogenomics of human AKRs needs to be studied with particular emphasis on single nucleotide polymorphisms. SNP's that result in amino-acid sequence changes can be mapped to the available crystal structures so that their effects on AKR structure-function can be predicted. Such polymorphisms may be responsible for individual susceptibility to drugs, toxicants and carcinogens and response to stress.

Acknowledgements

Supported by grants R01-CA39504 and P01-CA92537 to T.M.P.

References

1. Jez, J. M.; Flynn, T. G.; Penning, T. M. *Biochem. Pharmacol.* **1997**, *54*, 639-647.
2. Jez, J. M.; Bennett, M. J.; Schlegel, B. P.; Lewis, M.; Penning, T. M. *Biochem. J.* **1997**, *326*, 625-636.
3. Jez, J. M.; Penning, T. M. *Chemico-Biol. Inter.* **2001**, *130-132*, 499-525.
4. Kador, P. F.; Robinson, Jr.; W. G.; Kinoshita, J. H. *Annu. Rev. Pharmacol. Toxicol.* **1985**, *25*, 691-714.
5. Burcyznski, M. E.; Sridhar, G. R.; Palackal, N. T.; Penning, T. M.; *J. Biol. Chem.*, **2001**, *276* 2890-2897.
6. Atalla, A.; Maser, E. Chem. Biol. Inter., **2001**, *130-132* 737-748.
7. Smithgall, T.E.; Harvey, R.G.; Penning, T.M. *J. Biol. Chem.* **1986**, *261* 6184-6191.
8. Ellis, E. M.; Judah, D. J.; Neal, G. E.; Hayes, J. D. *Proc. Natl. Acad. Sci. USA*, **1993**, *90* 10350-10354.
9. Askonas, L. J.; Ricigliano, J. W.; Penning, T. M. *Biochem. J.* **1991**, *278* 835-841.
10. Kubiseski, T. J.; Hyndman, D. J.; Morjana, N. A.; Flynn, T. G. *J. Biol. Chem.* **1992**, *267*, 6510-6517.

11. Grimshaw, C. E.; Bohren, K. M.; Lai, C. J.; Gabbay, K. H.; *Biochemistry*, **1995**, *34,* 14356-14365.
12. Bohren, K. M.; Grimshaw, C. E.; Lai, C-J.; Harrison, D. H.; Ringe, D.; Petsko, G. A.; Gabbay, K. H. *Biochemistry* **1994**, *33,* 2021-2032.
13. Schlegel, B. P.; Jez, J. M.; Penning, T. M. *Biochemistry* **1998**, *37,* 3538-3548.
14. Hyndman, D.; Bauman, D. R.; Heredia, V.; Penning, T. M. *Chem. Biol. Inter.* **2003**, 143-144: 621-631.
15. Kozma, E.; Brown, E.; Ellis, E. M.; Lapthorn, A. J.; *J. Biol. Chem.* **2002**, *277,* 16285-16293.
16. Kelly, V. P.; Sherratt, P. J.; Crouch, D. H.; Hayes, J. D. *Biochem. J.* **2002**, In press.
17. Sato, S.; Kador, P. F. *Biochem. Pharmacol.* **1990**, *40,* 1033-1042.
18. Vander Jagt, D. L, Hassebrook, R.K., Hunsaker, L.A., Brown, W.M. Royer, R.E. *Chem. Biol. Inter.,* **2001**, *130-132* 549-562.
19. Gardner, H. W., *Free Rad. Biol. Med.* **1989**, *7* 65-86.
20. Porter, N. A.; Caldwell, S. E.; Mills, K. A.; *Lipids* **1995**, *30,* 277-290.
21. Uchida, K.; Toyokuni, S.; Nishikawa, K.; Kawakishi, S.; Oda, H.; Hiai, H.; Stadtman, E. R. *Biochemistry* **1994**, *33,* 12487-12494.
22. Yoritaka, A.; Hattori, N.; Uchida, K.; Tanaka, M.; Stadtman, E. R.; Mizuno, Y. *Proc. Natl. Acad. Sci.* USA, **1996**, *93* 2696-2701.
23. Markesbery, W. R.; Carney, J..M. *Brain Pathol.* **1999**, *9* 133-146.
24. Burczynski, M. E.; Lin, H-K.; Penning, T..M.; *Cancer Res.* **1999**, *59* 607-614.
25. Ramana, K V.; Dixit, B.L.; Srivastava, S.; Balendiran, G. K.; Srivastava, S.; Bhatnagar, A. *Biochemistry* **2000**, *39,* 12172-12180.
26. Srivastava, S.; Chnadra, A.; Wang, L-F.; Seifert Jr.; W. E.; DaGue, B. B.; Ansai, N. H.; Srivastava, S. K.; Bhatnagar, A. *J. Biol. Chem.* **1998**, *273,* 10893-10900.
27. Hecht, S. S.; *J. Natl. Cancer Institute* **1999**, *91,* 1194-1210.
28. Hecht, S. S.; *Chem. Res. Toxicol.* **1998**, *11,* 560-603.
29. Cavalieri, E. L.; Rogan, E. G. *Xenobiotica* **1995**, *25,* 677-688.
30. Gelboin, H. V. *Physiol. Rev.* **1980**, *60,* 1107-1166.
31. Conney, A. H., *Cancer Res.* **1982**, *42,* 4875-4917.
32. Buening, M. K.; Wilsocki, P. G.; Levin, W.; Yagi, H.; Thakker, D. R.; Akagi, H.; Koreeda, M.; Jerina, D. M.; Conney, A. H. *Proc. Natl. Acad. Sci.* USA **1978**, *75,* 5358-5361.
33. Chang, R. L.; Wood, A. W.; Conney, A. H.; Yagi, H.; Sayer, J. M.; Thakker, D. R.; Jerina, D. M.; Levin, W. *Proc. Natl. Acad. Sci.* USA, **1987**, *84,* 8633-8636.
34. Penning, T. M.; Burczynski, M. E.; Hung, C-F.; McCoull, K. D.; Palackal, N. T.; Tsuruda, L. S. *Chem. Res. Toxicol.* **1999**, *12,* 1-18.
35. Palackal, N. T.; Burczynski, M.E.; Harvey, R.G.; Penning, T. M. *Biochemistry,* **2001,** *40,* 10901-10910.

36. Palackal, N. T.; Lee, S-H.; Harvey, R. G.; Blair, I. A.; Penning, T. M. *J. Biol. Chem.* **2002**, *27*, 24799-24808.
37. Guengerich, F. P.; Arneson, K. O.; Willims, K. M.; Deng, Z.; Harris, T. M., *Chem. Res. Toxicol.* **2002** *15*, 780-792.
38. Kelly, V. P.; Ireland, L. S.; Ellis, E. M.; Hayes, J. D. *Biochem. J.* **2000**, *348*, 389-400.
39. Ireland, L. S.; Harrison, D. J.; Neal, G. E.; Hayes, J. D. *Biochem. J.* **1998**, *332*, 21-34.
40. Knight, L. P.; Primiano, T.; Groopman, J. D.; Kensler, T. W.; Sutter, T.R. *Carcinogenesis* **1999**, *20*, 1215-1223.
41. Ruepp, B.; Bohren, K. M.; Gabbay, K. H. *Proc. Natl. Acad. Sci.* USA, **1996**, *93*, 18318-18321.
42. Ferraris, J. D.; Williams, C. K.; Jung, K-Y.; Bedord, J. J.; Burg, M. B.; Garcia-Perez, A. *J. Biol. Chem.* **1996**, *271*, 18318-18321.
43. Rosette, C.; Karin, M. *Science* **1996**, *274*, 1194-1197.
44. Hung, C-F.; Penning, T. M. *Mol. Endocrinol.* **1999**, *13*, 1704-1717.
45. Miyakawa, H., Woo, S.K., Dahl, S.C., Handler, J.S., Kwon, H.M. *Proc. Natl. Acad.* USA **1999** *96*, 2538-2542.
46. Dahl, S.C.; Handler, J.S.; Kwon, H.M. *Am. J. Physiol.* **2001** *280* C248-C253.
47. Shinmura, K.; Bolli, R.; Liu, S-Q.; Tang, X-L.; Kodani, E.; Xuan, Y-L; Srivastasta, S.; Bhatnagar, A. *Cir. Res.* **2002** *91*: 240-246.
48. Angel, P.; Imagawa, M.; Chiu, R.; Stein, B.; Imbra, R. J.; Rahmsdorf, H. J.; Jonat, C.; Herrlich, P.; Karin, M. *Cell* **1987**, *49*, 729-739.
49. Angel, P.; Karin, M. *Biochimica. Biophysica Acta* **1991**, *1072*, 129-157.
50. Ciaccio, P.; Jaiswal, A. K.; Tew, K. D. *J. Biol. Chem.* **1994**, *269*, 15558-15562.
51. Itoh, K.; Chiba, T.; Takahashi, S.; Ishii, T.; Igarrashi, K.; Katoh, Y.; Oyake, T.; Hayasshi, N.; Satoh, K.; Hatayama, I.; Yamamoto, M.; Nabeeshima, Y-i, . *Biochem. Biophys. Res. Commun.* **1997**, *236*, 313-322.
52. Ciaccio, P. J., Stuart, J. E., Tew, K. D. *Mol. Pharmacol.* **1993**, *43*, 845-853.
53. Felsted, R. I.; Gee, M.; Bachur, N. R. *J. Biol. Chem.* **1974**, *259*, 3672-3679.
54. Penning, T. M. *Endocrine Rev.* **1997**, *18*, 281-305.
55. Matsuura, K.; Shiraishi, H.; Hara, A.; Sato, K.; Deyashiki, Y.; Ninomiya, M.; Sakai, S. *J. Biochem.* (Tokyo), **1998**, *124*, 940-946.

Aldo-Keto Reductases
and Endogenous Toxicants

Chapter 2

Aldo-Keto Reductase-Catalyzed Detoxication of Endogenous Aldehydes Associated with Diabetic Complications

David L. Vander Jagt, Lucy A. Hunsaker, Brian S. Young, and William M. Brown

Department of Biochemistry and Molecular Biology, University of New Mexico School of Medicine, Albuquerque, NM 87131

Numerous reactive aldehydes elevated in diabetic patients are key intermediates in the formation of Advanced Glycation Endproducts and likely contribute to development of long-term diabetic complications. These aldehydes include formaldehyde, glyoxal, methylglyoxal, glucosone, 3-deoxyglucosone, xylosone, 3-deoxyxylosone, and 4-hydroxynonenal. All of these aldehydes are substrates of human aldehyde and aldose reductases (AKR1A1, AKR1B1), two members of the aldo-keto reductase superfamily. The broad specificity of aldose and aldehyde reductases for these endogenous aldehydes suggests that detoxication of reactive aldehydes is one of their main functions. The structural features that contribute to substrate recognition include the presence of an oxidized carbon at the 2-position. Remaining structural features can vary widely. Thus, aldose and aldehyde reductases mainly appear to recognize a reactive aldehyde functional group. This imparts to these reductases an exceptionally broad protective function against the toxicity of reactive aldehydes.

A number of Advanced Glycation Endproducts (AGE) produced by modification of proteins by glucose have been identified either from *in vitro* or *in vivo* studies. These are thought to contribute to the cross-linking that is associated with the development of long-term diabetic complications. Although it initially was assumed that AGE are formed primarily from glycation of proteins by glucose, it is now clear that AGE can be formed from a variety of compounds besides glucose, including fructose, trioses, ribose, and ascorbate, and even from lipoxidation pathways (*1-14*). Therefore, it may be preferable to define AGE as Advanced Glycation/Lipoxidation Endproducts. In addition, reactive 2-oxoaldehydes appear to be key intermediates in the formation of many of the AGE identified thus far (Figure 1). The recent success in developing potential therapeutic agents that degrade AGE, points to the importance of AGE in the etiology of diabetic complications (*15-18*). The known AGE primarily are the products of reactions involving the 2-oxoaldehydes glyoxal, methylglyoxal and 3-deoxyglucosone.

The importance of methylglyoxal (MeG) deserves special attention: 1) Antibodies against MeG-derived AGE cross-react with AGE produced by modification of proteins with glucose, fructose, ribose, glyceraldehyde, glyoxal, ascorbate and ascorbate oxidation products, suggesting that MeG may be a common intermediate in AGE formation from a wide variety of glycating agents (*8*); 2) MeG-derived AGE and glyoxal-derived AGE as well as 3-deoxyglucosone-derived AGE are elevated in diabetes (*6-8*); 3) AGE derived from MeG and other 2-oxoaldehydes catalyze the production of free radicals (*19*); and 4) enzymes involved with metabolism of MeG are elevated in diabetes (*20*). These recent observations support the suggestion that MeG and other 2-oxoaldehydes play an essential role in the chemistry of AGE production.

Numerous studies support a role for aldose reductase (aldo-keto reductase, AKR1B1) in the development of diabetic complications in experimental models (*21-23*). A role for AKR1B1 in the development of diabetic complications in man is less certain. All of the endogenous aldehydes that have been implicated in AGE formation are substrates of AKR1B1; this includes MeG, glucose, glucosone, 3-deoxyglucosone, glyoxal, xylosone and 3-deoxyxylosone. The lipid-derived aldehyde 4-hydroxynonenal is also an excellent substrate of aldose reductase (*24-31*). This raises the question of the physiological roles of AKR1B1 and other aldo-keto reductases. In the present study, we compare the catalytic properties of AKR1B1 and the related aldo-keto reductase AKR1A1 (aldehyde reductase) with a range of reactive aldehydes and demonstrate that the broad specificities of AKR1A1 and AKR1B1 are consistent with the suggestion that a major function of these aldo-keto reductases is the detoxication of reactive aldehydes. We also demonstrate that glutathione plays a role in the detoxication of certain reactive aldehydes catalyzed by AKR1A1 and AKR1B1 and that there appears to be a glutathione binding site at the active sites of both AKR1A1 and AKR1B1.

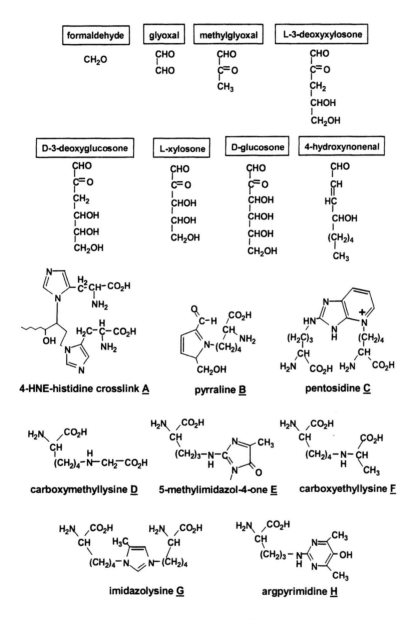

Figure 1: Structures of reactive aldehydes and AGEs implicated in the development of diabetic complications. AGEs E - H are derived from methylglyoxal.

Materials and Methods

Chemicals

MeG was prepared by hydrolysis of pyruvic aldehyde dimethylacetal with sulfuric acid followed by azeotropic distillation with water. 3-Deoxyglucosone, glucosone, 3-deoxyxylosone, xylosone, and 4-hydroxynonenal were synthesized by literature procedures. All other aldehydes were from Sigma.

Purification of AKR1A1 and AKR1B1

AKR1A1 and AKR1B1 were purified from human liver and skeletal muscle, respectively, as described previously (24-26).

Enzyme Assay

AKR1A1 and AKR1B1 activity in the direction of aldehyde reduction was routinely measured in 1 ml total volume of 0.1 M sodium phosphate buffer pH 7.0 with 10 mM glyceraldehyde and 0.1 mM NADPH. Enzyme activity was determined by following changes in NADPH concentration at 340 nm, $\varepsilon = 6.2$ $mM^{-1} cm^{-1}$.

Enzyme Kinetic Studies

Initial velocity studies were conducted in the buffer described above at 25 °C. Michaelis constants for substrates and cofactors and k_{cat} values were determined by nonlinear regression analysis of the initial rate data using the ENZFITTER program (Elsevier-Biosoft).

Flexible Docking

Flexible docking of glutathione to AKR1A1 and AKR1B1 was performed using the AutoDock 3.0 software suite from Scripps Research Institute (32). The crystal structure of human AKR1B1 (PDB accession1ADS) was modified to accommodate the docking (33). The coordinates of polar hydrogens were added as predicted by Sybyl 6.6 using torsional minimization. Partial charges were assigned from united Kollman dictionary charges and all substrate and ordered water atoms were removed. Inhibitor structure was predicted from GA conformational search followed by BFGS minimization in Sybyl. Inhibitor partial charges were assigned according to the Gasteiger-Huckel method. For human AKR1A1, the coordinates (PDB accession1CWN) of porcine aldehyde reductase (34) were used to construct a homology model of AKR1A1.

Results and Discussion

Substrate Properties of AKR1A1 and AKR1B1 with Aldehydes Implicated in the Development of Diabetic Complications

The catalytic efficiencies k_{cat}/K_m of AKR1A1 and AKR1B1 with the reactive aldehydes shown in Figure 1 are summarized in Table I. The k_{cat}/K_m values are 5-100 fold greater for AKR1B1 than for AKR1A1, with the highest difference observed with MeG, which appears to be the preferred substrate of AKR1B1. There is a marked increase in k_{cat}/K_m for any of the substrates compared to formaldehyde, raising the question of the role of the groups attached to the reactive aldehyde functional group. To address this question, the catalytic efficiencies of AKR1A1 and AKR1B1 were compared using a series of 2-carbon and 3-carbon aldehydes, summarized in Table II. For both the 2-carbon and 3-carbon series, k_{cat}/K_m values increase markedly if the 2-position of

**Table I. AKR1A1 and AKR1B1 – Catalyzed Reduction
of Reactive Aldehydes**

Substrate	k_{cat} / K_m (M^{-1} min^{-1})	
	AKR1A1	*AKR1B1*
formaldehyde	2.1×10^1	2.2×10^3
glyoxal	1.6×10^4	3.0×10^5
methylglyoxal	1.8×10^5	1.8×10^7
xylosone	1.1×10^5	5.3×10^5
3-deoxyxylosone	5.4×10^5	2.1×10^6
glucosone	8.5×10^4	5.5×10^5
3-deoxyglucosone	5.8×10^5	2.6×10^6
4-hydroxynonenal	$-$ [a]	4.6×10^6

Adapted in part from data in References 26, 28, 30.

[a] Low solubility precluded determination of kinetic parameters.

the aldehyde is an oxidized carbon. Thus, glycoaldehyde, glyoxal and glyoxylate are much better substrates than acetaldehyde, and acrolein, glyceraldehyde and MeG are much better substrates than propionaldehyde. This difference is even greater than suggested by the k_{cat}/K_m values when one considers that the aldehyde functional groups of these aldehydes are highly hydrated compared to acetaldehyde and propionaldehyde, leaving the concentration of free aldehyde that is the actual substrate of AKR1A1 and AKR1B1 very low. For example, MeG is 99.8% hydrated in solution (35). If corrected for this factor, the catalytic efficiency of AKR1B1 with MeG as substrate would approach a difussion limited value.

Table II. AKR1A1 and AKR1B1 -- Catalyzed Reduction of 2-Carbon and 3-Carbon Aldehydes

Substrate	k_{cat}/K_m $(M^{-1}\ min^{-1})$	
	AKR1A1	*AKR1B1*
CH_3-CH (O)	5.1×10^1	1.7×10^3
$HOCH_2-CH$ (O)	3.6×10^4	9.5×10^5
$HC-CH$ (O O)	1.6×10^4	3.0×10^5
HO_2C-CH (O)	3.3×10^5	6.7×10^4
CH_3-CH_2-CH (O)	2.2×10^2	6.5×10^4
$CH_2=CH-CH$ (O)	2.0×10^4	1.1×10^6
$HOCH_2-CH-CH$ (O) OH	1.6×10^5	7.5×10^6
CH_3-C-CH (O O)	1.8×10^5	1.8×10^7

The results in Tables I and II support the conclusion that AKR1A1 and AKR1B1 show broad specificities for reactive aldehydes that are highly hydrated and that these aldo-keto reductases appear to be designed to recognize the aldehyde functional group with limited sensitivities to the rest of the substrate, as long as the 2-position is an oxidized carbon. This preference for aldehydes with oxidized carbons at the 2-position can be ascribed in part to the H-bonding roles of residues H112 and H110 in AKR1A1 and AKR1B1, respectively, in dictating substrate specificity (36).

Substrate Specificity of Reduced and Oxidized AKR1B1

AKR1B1, unlike AKR1A1, has an active site cysteine residue (C298) that is sensitive to redox conditions. Simply dialyzing AKR1B1 in the absence of cofactor will produce an oxidized form that can be converted back into the reduced form by treatment with DTT (24,24). The substrate properties of oxidized AKR1B1 differ markedly from those of reduced AKR1B1, as shown in Table III. Additional forms of AKR1B1 involving nitrosation or gluta-thionylation of C298 can be formed, each with unique kinetic properties (37,38). In general, any modification of AKR1B1 will decrease k_{cat}/K_m values for any of the substrates. The physiological significance of the redox-sensing activity of AKR1B1 is not known. By comparison, AKR1A1 does not exhibit these properties, suggesting that AKR1A1 may be the more important of these reductases for detoxication of reactive aldehydes under conditions of oxidative stress that may compromise the activity of AKR1B1.

Table III. Substrate Properties of Reduced and Oxidized AKR1B1

Substrate	k_{cat}/K_m $(M^{-1}\ min^{-1})$	
	AKR1B1 (reduced)	AKR1B1 (oxidized)
glucose	9.1×10^2	$-$ [a]
glyceraldehyde	7.5×10^6	1.9×10^5
methylglyoxal	1.8×10^7	6.6×10^4

[a] Glucose is essentially not a substrate of oxidized AKR1B1.
Adapted in part from data in Reference 27.

Role of Glutathione in AKR1A1 and AKR1B1-Catalyzed Reactions

Glutathione reacts non-enzymatically with MeG to form the hemithioacetal. Thus intracellular MeG is an equilibrium mixture of free and hydrated MeG along with the glutathione-MeG hemithioacetal (Figure 2). AKR1B1 is able to catalyze the reduction of the aldehyde functional group of MeG to produce acetol as well as the ketone of the hemithioacetal to produce the hemithioacetal of lactaldehyde, which can then form lactaldehyde (*39*). Thus, with MeG as substrate AKR1B1 is both an aldehyde and a ketone reductase. Both lactaldehyde and acetol can be reduced further to propananediol, catalyzed by AKR1B1, again demonstrating the ability of AKR1B1 to function as a ketone reductase.

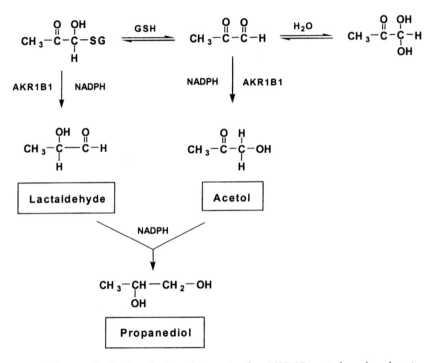

Figure 2. Role of glutathione in the AKR1B-catalyzed reduction of methylglyoxal (adapted from Reference 38).

The known ability (*37*) of AKR1B1 to catalyze the reduction of the Michael adducts of glutathione and α,β−unsaturated aldehydes is in agreement with the suggestion of a glutathione binding site at the active site of AKR1B1. The presence of a glutathione binding site near the active site of both AKR1A1 and AKR1B1 is supported by molecular modeling (Figure 3).

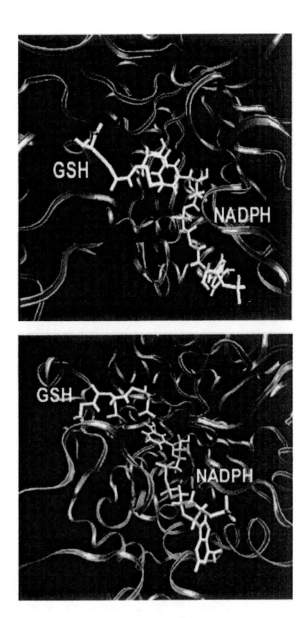

Figure 3. Molecular modeling identified GSH binding sites in AKR1A1 (top) and AKR1B1 (bottom).

AKR1A1 and AKR1B1-Catalyzed Reduction of Hydrophobic Aldehydes

The reduction of aldehydes catalyzed by AKR1A1 and AKR1B1 is not limited to small reactive aldehydes. Hydrophobic aldehydes, including aldehydes that do not possess an oxidized 2-carbon, can also be reduced efficiently (Table IV). In this case, the large substrate binding pocket of AKR1A1 and AKR1B1 provides the binding energy for substrate recognition (28,30).

Table IV. AKR1A1 and AKR1B1 --
Catalyzed Reduction of Hydrophobic Aldehydes

Substrate	k_{cat} / K_m (M^{-1} min^{-1})	
	AKR1A1	AKR1B1
	3.7×10^5	1.0×10^8
	2.1×10^6	1.1×10^8
	1.3×10^7	5.3×10^7

Adapted in part from data in Reference 28.

Summary

In summary, AKR1A1 and AKR1B1 exhibit broad specificities for reduction of aldehydes and in some cases ketones, derived from a versatile active site (Figure 4) that can accommodate reactive aldehydes with oxidized carbons at the 2-position, hydrophobic substrates, and selected glutathione adducts of aldehydes. This is consistent with the suggestion that a major function of AKR1A1 and AKR1B1 is detoxication of aldehydes.

Aldehyde Detoxication Functions of AKR1A1 and AKR1B1

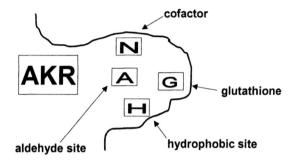

1) A site – small reactive aldehydes are substrates
2) N site – NADPH is the preferred cofactor
3) G site – glutathione adducts can be substrates
4) H site – hydrophobic aldehydes are substrates

Figure 4: Summary of the active site regions of AKR1A1 and AKR1B1 that provide broad ability to catalyze the reduction of aldehydes.

Acknowledgement

This work was supported by grants from the Juvenile Diabetes Foundation (1-1999-535) and National Institutes of Health (EY13695).

References

1. Bucala, R.; A. Cerami, A. *Adv. Pharmacol.* **1992,** 23, 1-34.
2. Cerami, A.; Stevens, V. J.; Monnier, V.M. *Metabolism* **1979,** 38, 431-438.
3. Brownlee, M. *Annual Rev. Med.* **1995,** 46, 223-234.
4. Fu,M-X.; Wells-Knecht, K. J.; Blackledge, J. A.; Lyons, T. J.; Thorpe, S. R; Baynes, J. W. *Diabetes* **1994,** 43, 676-683.
5. Grandee, S. K.; Monnier, V.M. *J. Biol. Chem.* **1991,** 266, 11649-11653.
6. Nagaraj, R. H.; Sell, D. R.; Prabhakaram, M.; Ortwerth, B. J.; Monnier, V.M. *Proc. Natl. Acad. Sci.* **1991,** 88, 10257-10261.
7. Reddy, S.; Bichler, J.; Wells-Knecht, K. J.; Thorpe, S. R.; Baynes, J. W. *Biochemistry.* **1991,** 34, 10872-10878.

8. Shamsi, F. A.; Partal, A.; Sady, C.; Glomb, M. A.; Nagaraj, R. H. *J. Biol. Chem.* **1998**, 273, 6928-6936.
9. Uchida, K.; Khor, O. T.; Oya, T.; Osawa, T.; Yasuda, Y.; Miyata, T. *FEBS Lett.* **1997**, 410, 313-318.
10. Hayase, F.; Konishi, Y.; Kato, H. (1995) *Biosci. Biotech. Biochem.* **1995**, 59, 1407-1411.
11. Lo, W. C.; Westwood, M. E.; McLellan, A. C.; Selwood, T.; Thornalley, P. J. *J. Biol. Chem.* **1994**, 269, 32299-32305.
12. Nagaraj, R. H.; Shipanova, I. N.; Faust, F. M. *J. Biol. Chem.* **1996**, 271,19338-19345.
13. Ahmed, M. U.; Brinkmann Frye, E.; Degenhardt, T. P.; Thorpe, S. R.; Baynes, J. W. *Biochem. J.* **1997**, 324, 565-570.
14. Yu, P. H. *J. Neural Transm.* Suppl. **1998**, 52, 201-216.
15. Brownlee, M.; Vlassara, H.; Kooney, A.; Ulrich, P.; Cerami, A. *Science* **1986**, 232, 1629-1632
16. Frank, R. N.; Amin, R.; Kennedy, A.; Hohman, T. C. *Arch. Ophthalmol.* **1997**, 115, 1036-1047.
17. Vasan, S.; Zhang, X.; Zhang, X.; Kapurhiotu, A.; Bernhagen, J.; Teichberg, S.; Basgen, J.; Wagle, D.; Shih, D.; Terlecky, I.; Bucala, R.; Cerami, A.; Egan, J.; Ulrich, P. *Nature* **1996**, 382, 275-278.
18. Wolffenbuttel, B. H.; Boulanger, C. M.; Crijns, F. R.; Huijberts, M. S.; Poitevin, P.; Swennen, G. N.; Vasan, S.; Egan, J. J.; Ulrich, P.; Cerami, A.; Levy, B. I. *Proc. Natl. Acad. Sci.* **1998**, 95, 4630-4634.
19. Lee, C.; Yim, M. B.; Chock, P. B.; Yim, H. S.; Kang, S. O. *J. Biol. Chem.* **1998**, 273, 25272-25278.
20. Ratliff, D.M.; Vander Jagt, D.J.; Eaton, R.P; Vander Jagt, D.L. *J. Clin. Endocrinol. Metab.* **1996**, 81, 488-492.
21. Gabbay, K. H.; Merola, L. O.; Field, R. A. *Science* **1966**, 151, 209-210.
22. Kador, P. F.; Kinoshita, J. H. *Am. J. Med.* **1985**, 79 (suppl 5A), 8-12.
23. Petrash, J. M.; Tarle, I.; Wilson, D. K.; Quiocho, F. A. *Diabetes* **1994**, 43, 955-959.
24. Vander Jagt, D. L.; Robinson, B.; Taylor, K. K.; Hunsaker, L. A. *J. Biol. Chem.* **1990**, 265, 20982-20987.
25. Vander Jagt, D. L.; Hunsaker, L. A.; Robinson, B.; Stangebye, L. A.; Deck, L. M. *J. Biol. Chem.* **1990**, 265, 10912-10918.
26. Vander Jagt, D. L.; Robinson, B.; Taylor, K. K; Hunsaker, L. A. *J. Biol. Chem.* **1992**, 267, 4364-4369.
27. Vander Jagt, D. L.; Hunsaker, L. A. *Adv. Exp. Med. Biol.* **1993**, 328, 279-288.
28. Vander Jagt, D. L.; Torres, J. E.; Hunsaker, L. A.; Deck, L. M.; Royer, R. E. *Adv. Exp. Med. Biol.* **1996**, 414, 491-497.

29. Kolb, N. S.; Hunsaker, L. A.; Vander Jagt, D. L. *Mol. Pharmacol.* **1994**, 45, 797-801.
30. Vander Jagt, D. L.; Kolb, N. S.; Vander Jagt, T. J.; Chino, J.; Martinez, F. J.; Hunsaker, L.A.; Royer, R.E. *Biochim. Biophys. Acta* **1995**, 1249, 117-126.
31. Vander Jagt, D. L.; Hunsaker, L. A.; Vander Jagt, T. J.; Gomez, M. S.; Gonzales, D. M.; Deck, L. M.; Royer, R. E. *Biochem. Pharmacology* **1997**, 53, 1133-1140.
33. Wilson, D. K.; Bohren, K. M.; Gabbay, K. H.; Quiocho, F. A. *Science* **1992**, 257, 81-84.
34. el-Kabbani, O.; Judge, K.; Ginell, S. L.; Myles, D. A.; DeLucas, L. J.; Flynn, T.G. *Nat. Struct. Biol.* **1995**, 2, 687-692.
35. Rae, C.; O'Donoghue, S. I.; Bubb, W. A.; Kuchel, P. W. *Biochemistry* **1994**, 33, 3548-3559.
36. Barski, O. A.; Gabbay, K. H.; Grimshaw, C. E.; Bohren, K. M. *Biochemistry* **1995**, 34, 11264-11275.
37. Ramana, K. T.; Dixit, B. L.; Srivastava, S.; Balendiran, G. K.; Srivastava, S. K.; Bhatnagar, A. *Biochemistry* **2000**, 39, 12172-12180.
38. Srivastava, S.; Dixit, B. L.; Ramana, K. V.; Chandra, A.; Chandra, D.; Zacarias, A.; Petrash, J. M.; Bhatnagar, A.; Srivastava, S. K. *Biochem. J.* **2001**, 58, 111-118.
39. Vander Jagt, D. L.; Hassebrook, R. K.; Hunsaker, L. A.; Brown, W. M.; Royer, R. E. *Chem Biol Interact.* **2001**, 130-132, 549-562.

Chapter 3

Aldose Reductase Detoxifies Lipid Aldehydes and Their Glutathione Conjugates

Satish K. Srivastava[1], Kota V. Ramana[1], Sanjay Srivastava[2], and Aruni Bhatnagar[2]

[1]Department of Human Biological Chemistry and Genetics, University of Texas Medical Branch, Galveston, TX 77555
[2]Division of Cardiology, Department of Medicine, University of Louisville, Louisville, KY 40202

Aldose reductase (AKR1B1, (AR)) is an NADPH dependent aldo-keto reductase (AKR) implicated in the pathogenesis of multiple secondary complications of diabetes. Treatment with AR inhibitors has been shown to prevent secondary complications of diabetes, however, the clinical efficiency of these drugs is limited due to generation of inhibitor-resistant forms of the enzyme in diabetic tissues. Our studies suggest that the development of inhibitor resistance is due to the induction of post-translational modifications of the enzyme mediated in part by concurrent changes in cellular glutathione and nitric oxide bioavailability. In addition, inhibition of the enzyme during diabetes could prevent the physiological functions of AR. Although reduction of glucose is believed to be the main physiological role of this enzyme, *in vitro* AR is a poor catalyst for glucose reduction and the hydrophobic active site of the enzyme is incompatible with hydrophilic substrates such as glucose. In contrast, AR is much more efficient in reducing hydrophobic short to medium chain aldehydes particularly α,β-unsaturated alkenals derived from lipid peroxidation such as 4-hydroxy *trans* 2-nonenal (HNE). The K_m of AR for HNE and related aldehydes is 10-30 μM, where as $K_{m \text{ glucose}}$ is 50-100 mM, suggesting that lipid-derived aldehydes may be the preferred substrates of the enzyme. Additionally, AR also reduces *in vivo* glutathiolated aldehydes with higher efficiency than their parent free aldehydes. Our kinetic, site-directed mutagenesis, and molecular modeling studies indicate that the AR active site is compatible with selective binding to glutathione in a specific orientation. Efficient reduction of hydrophobic aldehydes and their glutathione conjugates by AR suggests that the enzyme may be part of a detoxification mechanism for removing endogenous and xenobiotic aldehydes. A critical role of AR in preventing the toxicity of lipid peroxidation products is suggested by our recent observations that inhibition of this enzyme prevents smooth muscle cell mitogenesis and cytokine signaling-process associated with increased generation of reactive oxygen species and the induction of lipid peroxidation. Thus, post-translational changes in AR and its antioxidant role need to be considered in designing anti-AR interventions for treating diabetic complications.

Aldose reductase was initially purified as a glucose reducing protein from seminal vesicles (1-3). The enzyme was postulated to be part of the polyol pathway for synthesizing fructose from glucose. This pathway involves the reduction of glucose to sorbitol, followed by the oxidation of sorbitol to fructose catalyzed by sorbitol dehydrogenase (1,2). The first step of the pathway, which involves the reduction of glucose to sorbitol, was shown to be catalyzed by AR. This step is rate-limiting for the entire pathway and is unique in that it utilizes non-phosphorylated glucose; unlike other pathways of glucose metabolism (i.e., glycolysis and the pentose phosphate pathway) which utilize phosphorylated glucose. Due to the high affinity of hexokinase, glucose is mostly converted to glucose-6-phosphate and AR-dependent reduction represents only a fraction (>3%) of the total glucose utilization (3-5). However, during hyperglycemia and diabetes, when the cellular availability of glucose is increased, flux through the polyol pathway also increases dramatically (3). Although it could be viewed as a protective mechanism for removing excessive glucose from hyperglycemic tissues, continued glucose metabolism by the polylol pathway could incite tissue injury and dysfunction (3,6). Reduction of glucose by AR generates sorbitol, which if not effectively converted to fructose, could accumulate in tissues. Because membranes are mostly impermeable to sorbitol, sorbitol accumulation could impose osmotic stress, resulting in the ionic imbalance and the loss of glutathione (7,8). Additionally, the oxidation of NADPH during sorbitol synthesis could diminish the availability of reducing equivalents for antioxidant processes such as glutathione reductase, thereby contributing to generalized oxidative stress and a diminished capacity to reduce reactive oxygen species (9). Increased activity of the polyol pathway during hyperglycemia could also contribute to hyperglycemic injury by synthesizing fructose, which is the precursor of potent glycosylating agents. Thus, increased fructose formation and the elevated synthesis of glycosylating species could add to the generation of advanced glycosylating end products (AGE) and consequently to the biochemical abnormalities associated with AGE formation (3). In this scenario, inhibition of AR would interrupt the vicious cycle of osmotic overload, NADPH depletion, AGE formation, and oxidative stress. Indeed, the contribution of AR to the etiology of hyperglycemic injury and the development of secondary diabetic complications was realized within 10 years of discovery of the enzyme and a number of inhibitors were synthesized to inhibit AR and to prevent tissue injury and dysfunction associated with several diabetic complications (6,10,11).

Extensive experimental data suggest that inhibition of AR could prevent, delay or in some cases even reverse tissue injury associated with the development of diabetic complications (10,11). Salutary and beneficial effects of these drugs against nephropathy, neuropathy, and cataractogenesis have been reported, particularly in the rat model of streptozotocin-induced diabetes (3). Nonetheless, clinical trials of AR inhibitors have been inconclusive, and the results of these studies do not constitute a firm demonstration of the efficacy of AR inhibitors in uniformly preventing human diabetic complications (10-12). Reasons for the disappointing clinical results with AR inhibitors have been debated in the literature for several years, however no clear consensus has emerged. Several of the earlier trials with sorbinil were marred by hypersensitivity reactions in an unacceptably high proportion of the tested population. In later trials, issues relating to the non-selectivity of the drugs, their inappropriate dosing and pharmacokinetics and high general toxicity of these compounds were raised but not resolved. As a result, the utility of AR inhibitors and the validity of anti-AR interventions in preventing and treating secondary diabetic complications remain unsubstantiated. Our kinetic and structural studies suggest additional reasons for the inconclusive results of clinical trials with AR inhibitors. On one hand these data show that the enzyme is somehow modified during diabetes and therefore is less susceptible to inhibition and on the other that AR plays a critical antioxidant role, and if inhibited could, in some circumstances, lead to worsening hyperglycemic injury.

Diabetic Changes in AR

Changes in the susceptibility of AR to inhibition were apparent from earlier studies showing that the enzyme from diabetic tissues displayed remarkably different kinetic properties than the protein isolated from non-diabetic tissues (13-15). Significantly, the diabetic enzyme was less sensitive to inhibition by sorbinil than the euglycemic form of the enzyme. Although structural reasons for the changes in the AR protein were not identified, these data clearly point to a post-translational change in the enzyme, and are consistent with evidence derived from cell culture studies showing that sorbitol formation in endothelial cells cultured with high glucose becomes progressively resistant to sorbinil *(16)*. To identify mechanisms and determinants of inhibitor binding to AR, extensive kinetic studies were performed with a series of AR inhibitors. These studies revealed that most of the AR inhibitors bind to the active site of the protein and not to a previously postulated inhibitor-binding domain independent of the substrate-binding site *(17,18)*. More significantly, perhaps these data demonstrated a critical role of protein thiols in inhibitor and substrate binding to the enzyme. It was found that whereas the reduced enzyme displayed high affinity for inhibitors such as sorbinil and alrestatin, oxidation of the enzyme, either inadvertently due to prolonged storage *(19)* or deliberately by thiol-reactive reagents *(20)* caused a dramatic decrease in the affinity of the enzyme for some, but not all, inhibitors. These changes in the affinity of the enzyme for sorbinil and related inhibitors were reversible upon reduction of the protein thiols *(19)*, which restored the loss of sensitivity caused by storage and oxidation or both.

The role of thiol oxidation in regulating the inhibitor sensitivity of the enzyme was substantiated by the observations that the enzyme purified in the presence of its cofactor NADPH, which binds to the active site of the protein, and prevents solvent access to the active site, preserved inhibitor sensitivity of the enzyme *(21)*. Because kinetic studies showed inhibitor binding overlapped with substrate binding at the active site, it became clear that the thiol(s) regulating the inhibitor-sensitivity of the enzyme must be located at the active site of the enzyme. By selective carboxymethylation with radioactive iodoacetate, Cys-298 was identified to be the main residue regulating the redox changes in the kinetic properties of AR and its substrate and inhibitor binding *(21,22)*. This role of Cys-298 was confirmed by site-directed mutagenesis studies showing that replacement of this residue by serine generated an enzyme, which displayed low affinity for substrates and inhibitors *(23)*, and was remarkably similar, at least in its kinetic behavior, to the enzyme purified from diabetic tissues. Whether oxidation of Cys-298 is the key post-translational modification of AR in diabetic tissues remains to be established. However, recent evidence suggests that this residue may be a unique target of the pluripotent regulatory molecule nitric oxide (NO) and that changes in NO during diabetes may be a significant factor in regulating the inhibitor-sensitivity of AR in diabetic tissues *(24)*.

Due to its high reactivity, Cys-298 of AR is highly sensitive to a variety of oxidants including NO and nitrosothiols. Exposure of the enzyme *in vitro* to NO-donors leads to inactivation *(25)*, and in case of nitrosothiols, to the formation of a mixed disulfide between the protein and the NO-bearing thiol *(26)*. That NO or nitrosothiol-mediated modification may be relevant to the *in vivo* situation is supported by the observations that exposure to an NO-donor inhibits sorbitol accumulation in erythrocytes from normal and diabetic rats

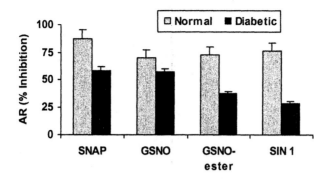

Figure 1. Nitric oxide donors prevent sorbitol formation in diabetic and non-diabetic rat erythrocytes. S-Nitroso-N-acetylpenicillamine (SNAP), S-nitrosoglutathione mono-ethyl-ester (GSNO-Ester), 3-morpholinosydnonimine (SIN-1).

(Figure 1), indicating that increased availability of NO inhibits AR and the flux of glucose through the polyol pathway *(27)*. Furthermore, exposure to NO donors or nitroglycerine patch prevents the accumulation of sorbitol in several tissues of diabetic rats and inhibition of nitric oxide synthesis activates AR and promotes sorbitol accumulation (data not shown). Taken together, these data suggest that the increase in the polyol pathway activity during diabetes is not a linear function of glucose availability, but is critically regulated by NO. Under euglycemic conditions, associated with high NO generation, the enzyme is repressed and inhibited, whereas during hyperglycemia, due to a decrease in NO bioavailability, AR is de-repressed and is activated such that it generates more sorbitol and contributes to the development of diabetic complications.

In contrast to its regulation of AR activity, the regulation of the AR inhibitor sensitivity by NO is more complex. *In vitro* studies show that, depending upon the nature of the NO donor, NO could either inhibit or activate AR *(28)*. When incubated with AR, *S*-nitrosoglutathione (GSNO) causes rapid inhibition of the enzyme *(26)*. Most of the activity is lost within 30 min and the inactivation reaches a steady-state. However, even at steady-state more than 20% of the enzyme activity remains uninhibited, and becomes resistant to inhibition by sorbinil. Structural analyses using electrospray mass spectrometry and site-directed mutagenesis revealed that inhibition of AR by GSNO is due to the formation of a single mixed disulfide between glutathione and Cys-298. Because Cys-298 is at the active site, binding of the bulky glutathione molecule at this site blocks substrate access and prevents catalysis *(26)*. A dramatically different modification is observed when AR is treated with reagents that cause protein S-nitrosylation. When incubated with NO donors such as SNAP and NONOate the maximal AR activity is increased several fold (28). This change appears also to be due to the modification of Cys-298, but in this case it is linked to the formation of an S-nitrosothiol rather than a mixed disulfide (24). Because S-nitrosylation introduces only a small molecule (NO) at the active site cysteine, the enzyme remains active and substrate access is not prevented. Indeed, the maximal activity of the S-nitrosylated enzyme is higher than that of the unmodified enzyme; a consequence perhaps of an increase in the rate of NADP dissociation, which is the rate-limiting step in the AR catalytic cycle and is increased by the oxidation of Cys-298. Significantly, the S-nitrosylated AR is insensitive to inhibitors such as sorbinil.

The view that emerges from these data is that modification of AR by NO-donors is complex and could lead to either activation or inhibition depending upon the nature of the modification induced in the protein. Although the *in vivo* modifications in AR protein induced by NO have not been identified, the *in vitro* data support the following speculative scheme of sequencial events during euglycemic and hyperglycemic conditions (Figure 2).

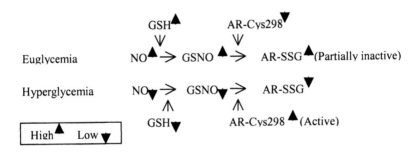

Figure 2. Regulation of AR activity by NO under euglycemic and hyperglycemic conditions.

Under euglycemic conditions, with high NO availability and high concentrations of reduced glutathione, the AR activity is partially inhibited due to NO-induced modification. In this case NO does not directly modify AR, but generates GSNO, which, in turn, forms a mixed disulfide with AR. That AR remains partially inhibited in euglycemic tissues by an NO-dependent mechanism is supported by our observation that inhibiting NO synthesis in normal rodents results in increased sorbitol accumulation (24). However, most of the enzyme is free and displays kinetic properties similar to the enzyme completely reduced *in vitro*, and it remains highly sensitive to inhibition by sorbinil or tolrestat. The constitutive repression of AR could represent a mechanism of keeping the polyol pathway in check when glucose availability is limited.

During diabetes, however, NO bioavailability is decreased *(29)*. Even if NO synthesis *per se* is not affected by diabetes, increased oxidative stress, particularly the increased formation of superoxide anions removes active NO by forming peroxynitrile, and glutathione is depleted *(30)*. These conditions (lack of glutathione and NO) derepress AR such that it becomes active. This scheme of events is supported by our observation that even at the same glucose concentration diabetic tissues synthesize more sorbitol than non-diabetic tissues (data not shown). This synthesis of sorbitol could be partially inhibited by feeding the NO precursor L-arginine, suggesting that activation of the polyol pathway during diabetes is due to de-repression of AR. Nonetheless, due to glutathione depletion and the formation of potent nitrosating species during diabetes, AR is modified by nitrosating species other than GSNO, leading to the generation of *S*-nitrosylated protein, which has a higher catalytic rate and is relatively insensitive to inhibitors such as sorbinil.

In its entirety, the postulated series of events suggest that diabetes diminishes the inhibitor sensitivity of AR, and that these changes are mediated by concurrent changes in glutathione and NO. Thus, clinical treatment of long-term diabetics with AR inhibitors may not translate into dramatic protection seen in short-term animal studies. Hence, diabetic changes in AR protein need to be considered when re-designing anti-AR interventions. Moreover, the data obtained from our studies on NO-modification of AR suggest that in addition to its other multifarious roles in neurotransmission, blood pressure regulation, or platelet aggregation, NO is an important regulator of the polyol pathway of glucose metabolism.

AR and Antioxidant Protection

An additional reason for the marginal efficacy of AR inhibitors in preventing the clinical symptoms of secondary diabetic complications may be that inhibition of the enzyme increases oxidative stress by preventing the detoxification of aldehydes derived from lipid peroxidation. That AR is involved in antioxidant defense was suggested by the initial observations that the lipid peroxidation product – 4-hydroxynonenal (HNE) and its glutathione conjugate are particularly good substrates of the enzyme *in vitro (31,32)*. It was found that in contrast to glucose, AR catalyzes the reduction of HNE and GS-HNE with a 1000-fold higher catalytic efficiency. Aldehydes such as HNE are generated at high concentrations by lipid peroxidation reactions. Because of α,β–unsaturation, these aldehydes are highly toxic and cause glutathione depletion, inhibition of respiration, changes in membrane excitability, formation of protein and DNA adducts and dysregulation of ion homeostasis *(33)*. Thus, reduction of these aldehydes by AR to inert alcohols could be an important mechanism for protecting against the toxicity of lipid peroxidation *(34-36)*.

Structural and kinetic studies show that the high hydrophobicity of the AR active site is compatible with the binding of short and medium chain aldehydes, such as those derived from lipid peroxidation. In contrast, the lack of ionic residues at the AR active site offers little opportunity for efficient binding of hydrophilic molecules such as glucose. The AR active site is lined by highly hydrophobic amino acid residues such as Trp20, Val47, Trp111, Phe121, Phe122, Pro218, Trp219, Leu300 and Leu301, and it binds efficiently to apolar aldehydes, particularly the C4 to C10 α,β-unsaturated aldehydes. Results shown in Figure 3 demonstrate that AR efficiently catalyzes the reduction of saturated and unsaturated aldehydes. As compared to short chain alkanals such as propanal, the long-chain aldehydes are reduced more efficiently. The presence of an additional methylene group in butanal, increases catalytic efficiency as compared to the propanal. Similarly, the presence of additional methylene groups in pentanal and hexanal, increases the catalytic efficiency of AR. However, the presence of additional methylene groups from heptanal to decenal gradually decreased the catalytic efficiency, whereas, these were better substrates than C3 and C4 saturated aldehydes. As with saturated aldehydes, the C4 to C10 unsaturated aldehydes are also reduced by the enzyme with increasing efficiency.

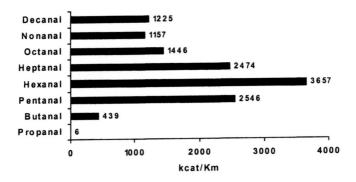

Figure 3. Catalytic efficiency (k_{cat}/K_m; min^{-1}/mM^{-1}) of AR for the reduction of saturated aldehydes.

In most cells, unsaturated aldehydes are metabolized by multiple metabolic pathways of which glutathione-S-transferase (GST)-catalyzed conjugation appears to be a central detoxification event. In addition to conjugation, these aldehydes could be either oxidized to their corresponding acids or reduced to alcohols. To examine the contribution of these pathways, we have studied the metabolism of HNE in several cardiovascular tissues including the heart, the endothelial and smooth muscle cells, and erythrocytes *(37-39)*. These studies reveal that although glutathiolation is a major metabolic fate of HNE, the conjugate of HNE is further reduced to the corresponding alcohol i.e., glutathionyl-1,4-dihydroxynonene (GS-DHN). Treatment with sorbinil or tolrestat prevents this reduction, indicating that at least in the cardiovascular tissues, glutathione conjugates are reduced by AR. The efficacy of AR in catalyzing the reduction of GS-HNE is substantiated by *in vitro* studies showing that the enzyme is particularly efficient in reducing glutathiolated aldehydes *(35,36)*.

The greater efficiency of AR towards glutathiolated aldehydes suggests selective interaction of active site amino acid residues with the glutathione backbone of the conjugates. Our molecular modeling studies suggest that glutathiolated aldehydes such as GS-propanal could potentially bind to the AR active site in two distinct orientations *(35,36)*. In orientation 1, the N-terminal glutamate forms one prong of the Y-shaped molecule parallel to the bound aldehyde. In this orientation, the carboxylate of γ-Glu forms H-bonds with Trp20-N, Lys21-N and its amino group H-bonds with Val47-O, whereas the carboxyl of the C-terminal glycine forms H-bonds with Leu-301-N, and Ser-302-N. In orientation 2, which is a $180°$ flip of the molecule along the aldehyde axis, Gly-3 is parallel to the aldehyde. In this orientation the carboxylate of the Glu-1 is H-bonded to Leu-301-N and Ser-302-N. Site-directed mutagenesis studies and structure-activity relationships suggest that the orientation 1 is the preferred mode of binding since replacement of Trp20 with phenylalanine disrupted the interaction of the protein with the γ-Glu of glutathione. This was also supported by energy-minimized models, which demonstrate that binding in orientation 1 is energetically favorable. Collectively, these studies demonstrate that the α/β-barrel structure of AR provides efficient scaffolding for glutathione binding comparable to that observed with other glutathione-dependent enzymes. Moreover, efficient binding and reduction of glutathione conjugates by AR supports the notion that reduction of glutathione conjugates is a significant feature of the physiological functions of this enzyme (Figure 4), and that inhibition of AR during diabetes could interfere with the detoxification of endogenous and xenobiotic aldehydes.

Role of AR in Cell Growth and Survival

In addition to metabolism and detoxification of toxic aldehydes and their glutathione conjugates, recent evidence suggests that AR may also be involved in regulating mitogenic signaling and cell growth *(40)*. Moreover, given the extensive evidence documenting the role of reactive oxygen signaling in cell growth, the mitogenic and the antioxidant roles of AR may be linked.

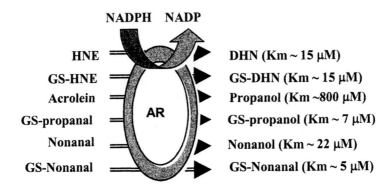

Figure 4. Relative K_m values for the reduction of free aldehyde and their glutathione conjugates by AR.

The role of AR in regulating cell growth was initially indicated by our studies showing that stimulation with growth factors and mitogens leads to an increase in the expression of AR (*41*) suggesting that upregulation of AR may be part of the delayed-early mitogenic responses. Our studies show that treatment with AR inhibitors prevents smooth muscle cell (SMC) growth in culture and that treatment with AR inhibitors decreases intimal hyperplasia in restenotic vessels (*41*). These observations suggest that AR is an obligatory mediator of cell growth. Our recent observations, demonstrating that inhibition of AR also prevents TNF-α-induced VSMC growth (see Ramana *et al.*, in this book) further extends and substantiates the mitogenic role of AR. Although the mechanisms by which inhibition of AR prevents cell growth remain to be fully elucidated, the observation that inhibition of AR abrogates PKC/IκB-α/NF-κB signaling pathway (*40*) suggests that the catalytic activity of AR is essential for the uninterrupted progress of a central pathway regulating cell growth and survival.

References

1. Clements, R. S. Jr. *Drugs* **1986**, 32 Suppl 2, 3-5.
2. Kinoshita, J. H. *Exp. Eye Res.* **1990**, 50, 567-573.
3. Bhatnagar, A.; Srivastava, S.K. *Biochem. Med. Metabol. Biol.* **1992**, 48, 91-121.

4. Johnson, E. C.; Young, M. K.; Stacy, P. A.; Beatty, C. H. *Clin. Chim. Acta* **1979**, 98, 77-85.
5. Cheng, H. M.; Xiong, J.; Tanaka, G.; Chang, C.; Asterlin, A. A.; Aguayo, J. B. *Exp. Eye Res.* **1991**, 53, 363-366.
6. Yabe-Nishimura, C. *Pharmacol. Rev.* **1998**, 50, 21-33.
7. Lou, M. F.; Dickerson, J. E., Jr; Garadi, R.; York, B. M., Jr. *Exp. Eye Res.* **1988**, 46, 517-30.
8. Ciuchi, E.; Odetti, P.; Prando, R. *Metabolism* **1996**, 45, 611-613.
9. Lee, A.Y.; Chung, S.S. *FASEB J* **1999**, 13, 23-30.
10. Tomlinson, D. R.; Stevens, E. J.; Diemel, L. T. *Trends in Pharmacol. Sci.* **1994**, 15, 293-297.
11. Costantino, L.; Rastelli, G.; Vianello, P.; Cignarella, G.; Barlocco, D. *Med Res Rev* **1999**, 19, 3-23.
12. Pfeifer, M. A.; Schumer, M. P. *Diabetes* **1995**, 44, 1355-1361.
13. Srivastava, S. K.; Hair, G. A.; Das, B. *Proc. Natl. Acad. Sci. USA* **1985** 82, 7222-72226.
14. Das, B.; Srivastava, S. K. *Diabetes* **1985**, 34, 1145-1151.
15. Srivastava, S. K.; Petrash, J. M.; Sadana, I. J.; Ansari, N. H.; Partridge, C. A. *Curr. Eye Res.* **1983**, 2, 407-410.
16. Lorenzi, M.; Toledo, S.; Boss, G. R.; Lane, M. J; Montisano, D.F. **1987** *Diabetologia 30*, 222-227.
17. Bhatnagar, A.; Liu, S-Q.; Das, B.; Ansari, N. H.; Srivastava, S. K. *Biochem. Pharmacol.* **1990**, 39, 1115-1124.
18. Liu, S-Q.; Bhatnagar, A.; Srivastava, S. K. *Biochem. Pharmacol.* **1992**, 44, 2427-2429.
19. Bhatnagar, A.; Liu, S.; Das, B.; Srivastava, S. K. *Mol. Pharmacol.* **1989**, 36, 825-830.
20. Liu, S-Q.; Bhatnagar, A.; Srivastava, S. K. *Biochim. Biophys. Acta* **1992**, *1120*, 329-336.
21. Vander Jagt, D. L.; Hunsaker, L. A.; Robinson, B.; Stangebye L. A.; Deck, L. M. *J. Biol. Chem.* **1990,** *265,* 10912-10918.
22. Liu, S. Q.; Bhatnagar, A.; Ansari, N. H.; Srivastava, S.K. *Biochim. Biophys. Acta* **1993**, 1164, 268-272.
23. Petrash, J. M.; Harter, T. M.; Devine, C. S.; Olins, P. O.; Bhatnagar, A.; Liu, S.; Srivastava, S. K. *J. Biol. Chem.* **1992**, 267, 24833-24840.
24. Chandra, D.; Jackson, E. B.; Ramana, K. V.; Srivastava, S. K.; Bhatnagar, A., *Diabetes* 2002, In Press.
25. Chandra, A.; Srivastava, S.; Petrash, J. M.; Bhatnagar, A.; Srivastava, S. K. *Biochim. Biophys. Acta* **1997**, *1341*, 217-222.
26. Chandra, A.; Srivastava, S.; Petrash, J. M.; Bhatnagar, A.; Srivastava S. K. *Biochemistry* **1997**, 36, 15801-15809.

48

27. Dixit, B. L.; Ramana, K. V.; Chandra, D.; Jackson, E. B.; Srivastava, S.; Bhatnagar, A.; Srivastava S. K. *Chem Biol Interact.* **2001**, 130-132, 563-571.
28. Srivastava, S.; Dixit, B. L.; Ramana, K. V.; Chandra, A.; Chandra, D.; Zacarias, A.; Bhatnagar, A.; Srivastava, S.K. *Biochem. J.* **2001**, 358, 111-118.
28. Lin, K. Y.; Ito, A.; Asagami, T.; Tsao, P. S.; Adimoolam, S.; Kimoto, M.; Tsuji, H.; Reaven, G. M.; Cooke, J. P. *Circulation* **2002**, 106, 987-992.
29. Ishii, N.; Patel, K. P.; Lane, P. H.; Taylor, T.; Bian, K.; Murad, F.; Pollock, J.S.; Carmines, P. K. *J. Am. Soc. Nephrol.* **2001** 12, 1630-1639.
30. Vander Jagt, D. L.; Kolb, N. S.; Vander Jagt, T. J.; Chino, J.; Martinez F. J.; Hunsaker, L. A.; Royer, R.E. *Biochim. Biophys. Acta.* **1995**, *1249*, 117-126.
31. Srivastava, S.; Chandra, A.; Bhatnagar, A.; Srivastava, S. K.; Ansari, N. H. *Biochem. Biophys. Res. Commun.* **1995**, 217, 741-746.
32. Esterbauer, H.; Schaur, R. J.; Zollner, H. *Free Radic. Biol. Med.* **1991**, *11*, 81-128.
33. Srivastava, S.; Watowich, S. J.; Petrash, J. M.; Srivastava, S. K.; Bhatnagar, A. *Biochemistry* **1999**, 38, 42-54.
34. Ramana, K. V.; Dixit, B. L.; Srivastava, S.; Balendirnan, G. K.; Srivastava, S. K. *Biochemistry* **2000**, 39, 12172-12180.
35. Dixit, B. L.; Balendiran, G. K.; Watowich, S. J.; Srivastava, S.; Ramana, K. V.; Petrash, J. M.; Bhatnagar, A.; Srivastava, S. K. *J. Biol. Chem.* **2000**, 275, 21587-21595.
36. Srivastava, S.; Chandra, A.; Wang, L. F.; Seifert, W. E., Jr; DaGue, B. B.; Ansari, N. H.; Srivastava, S. K.; Bhatnagar, A. *J. Biol. Chem.* **1998,** 273, 10893-10900.
37. Srivastava, S.; Conklin, D. J.; Liu, S. Q.; Prakash, N.; Boor, P. J.; Srivastava, S. K.; Bhatnagar A. *Atherosclerosis* **2001**, 158, 339-350.
38. Srivastava, S.; Dixit, B. L.; Cai, J.; Sharma, S.; Hurst, H. E.; Bhatnagar, A.; Srivastava, S. K. *Free Radic. Biol. Med.* **2000,** 29, 642-651.
39. Ramana, K. V.; Chandra, D.; Srivastava, S.; Bhatnagar, A.; Aggarwal, B. B.; Srivastava, S. K. *J. Biol. Chem.* **2002**, 277, 32063-32070.
40. Ruef, J.; Liu, S. Q.; Bode, C.; Tocchi, M.; Srivastava, S.; Runge, M. S.; Bhatnagar, A. *Arterioscler Thromb. Vasc. Biol.* **2000**, *20*, 1745-1752.

Chapter 4

Role of Aldose Reductase in the Detoxification of Oxidized Phospholipids

Sanjay Srivastava[1], Kota V. Ramana[2], Satish K. Srivastava[2], Stanley E. D'Souza[3], and Aruni Bhatnagar[1]

Departments of [1]Medicine, Division of Cardiology and [3]Physiology and Biophysics, University of Louisville, Louisville, KY 40202
[2]Department of Human Biological Chemistry and Genetics, University of Texas Medical Branch, Galveston, TX 77555

Oxidation of phospholipids generates products in which unsaturated fatty acids at the sn-2 position are oxidized to short chain aldehydes or epoxides. Such phospholipids containing phosphatidylcholine display biological activity similar to platelet activating factor (PAF) and stimulate monocyte adhesion to endothelial cells. These lipids can also serve as recognition sites for scavenger receptors. However, the mechanisms by which cells metabolize and detoxify oxidized phospholipids are not known. We therefore examined the possibility that oxidized phospholipids are removed by metabolism via aldose reductase (AR). *In vitro* human recombinant AR was found to be an excellent catalyst for the reduction of 1-palmitoyl-2-(5-oxovaleroyl)-phosphatidyl choline (POVPC) and related short to medium chain aldehydes that are likely to be derived from POVPC by phospholipase A_2-mediated metabolism. These observations suggest that metabolism via AR could be a potential mechanism for the detoxification of oxidized phospholipids generated during pathological oxidative stress.

Oxidative stress has been implicated in the etiology of several cardiovascular diseases including atherosclerosis, ischemia-reperfusion, and heart failure. Extensive investigations have shown that the pivotal step in the development of atherosclerosis is the oxidation of low-density lipoproteins (LDL) *(1,2)*. The oxidized LDL is the major source of vascular injury, and components of oxidized LDL stimulate the binding of leukocytes to the vessel wall and trigger the uptake of the lipoprotein particle by macrophages, smooth muscle cells, and the endothelium. Continued generation of oxidized products in LDL induces the generation of cytokines, promotes smooth muscle cell growth, and endothelial cell apoptosis in the endothelial cells. Nonetheless, the components of LDL that trigger and sustain the atherosclerotic sequelae have not been identified and their contribution to specific stages of atherosclerosis has not been assessed.

LDL oxidation generates an array of bioactive molecules of which aldehydes are a major component. Oxidation of the lipids is initiated by proton abstraction from a double bond of unsaturated fatty acids, leading to the generation of peroxyl radicals and peroxides. The reaction is propagated by dismutation of peroxides to alkoxyl and peroxyl radicals. Due to the presence of multiple *bis*-allylic methylenes in polyunsaturated fatty acids, a large number of products are generated, of which aldehydes comprise a large fraction *(3)*. One likely mechanism of aldehyde generation from oxidative C-C cleavage of cyclized hydroperoxide intermediates is shown in Figure 1. Similar reactions give rise to other carbonyl products, and additional reactions also generate aldehydes *(3,4)*. The free aldehyde generated could be of different chain length, depending upon exact site of cleavage. The chain length of the aldehyde remaining esterified could also vary. Thus, free and esterified aldehydes with a different number of double bonds and methylene chains are formed. Among free unsaturated aldehydes generated from the peroxidation of ω-6-polyunsaturated fatty acids (arachidonate, linolinate and linolenate), 4-hydroxy *trans*-2-nonenal (HNE) is formed in the highest concentration, and under some conditions, accounts for > 95% of alkenals produced *(5)*. Normal cellular levels of HNE are in the range of 0.8 to 2.8 μM *(4)*. However, under the condition of oxidative stress, the concentration of HNE in the membrane can reach up to 5 mM and the concentration of HNE in oxLDL has been estimated to be 150 mM *(4)*.

Lipid-derived aldehydes such as HNE are highly reactive *(4)* and lead to apoptosis *(6)* or necrosis *(7)*. However, in sub-lethal concentrations, they profoundly affect cell metabolism and signaling. Submicromolar concentrations of these aldehydes have been shown to induce permeability transition in isolated mitochondria *(8)*. In addition, HNE activates stress responsive protein kinases such as c-jun terminal kinase (JNK) and p38, and increases the cellular production of reactive oxygen species *(9)*. In addition, low concentrations of HNE sustain a proinflammatory state, by inhibiting the NFκ-B/Rel system *(10)*, and the ubiquitin/proteasome-dependent proteolysis *(11)*.

The core aldehydes, such as 1-palmitoyl-2-(5-oxovaleroyl) phosphatidylcholine (POVPC), are derived from the oxidation of 1-palmitoyl-2-arachidonyl-glycero-3-phosphocholine (PAPC), and represent the β-scission

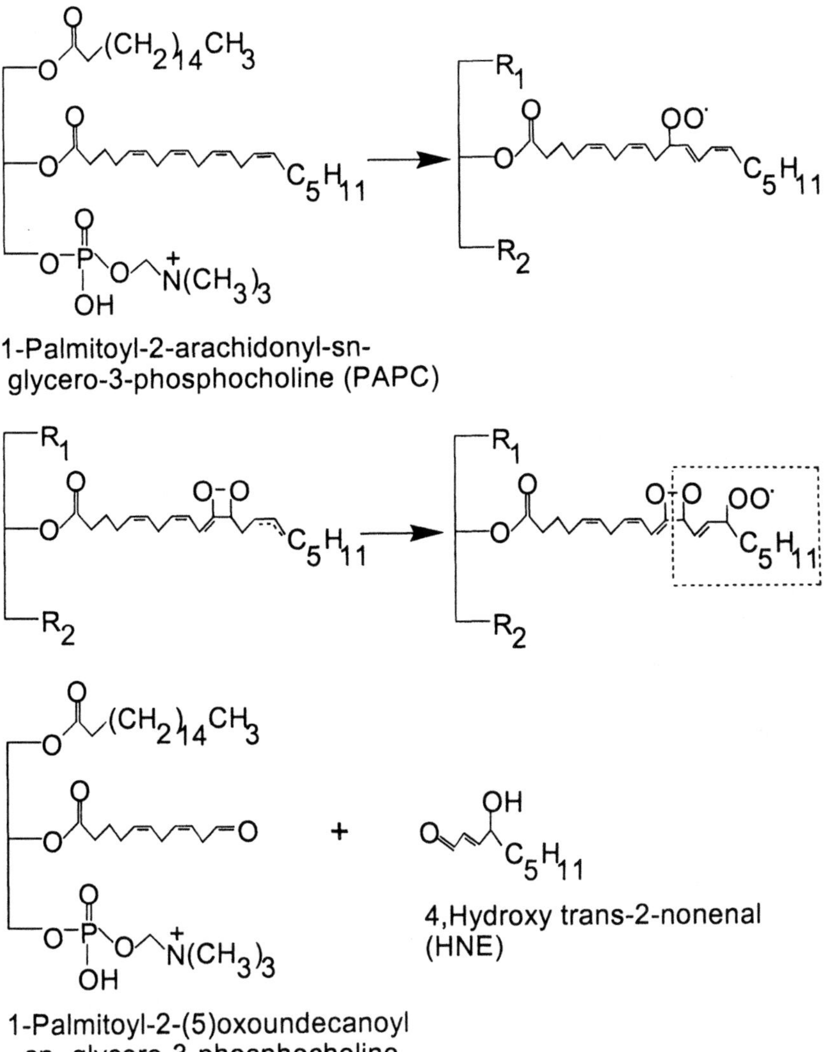
(CH2)14CH3
C5H11
OO·
R1
R2
N(CH3)3
OH
1-Palmitoyl-2-arachidonyl-sn-
glycero-3-phosphocholine (PAPC)
O-O
OH
4,Hydroxy trans-2-nonenal
(HNE)
1-Palmitoyl-2-(5)oxoundecanoyl
-sn- glycero-3-phosphocholine

Figure 1. Generation of aldehydes during phospholipid oxidation.

product derived from the inner most *bis*-allylic double bond of the arachidonyl chain *(12)*. POVPC is generated in high concentrations in minimally oxidized LDL (mmLDL) and in fatty streak lesions of cholesterol-fed rabbits. It has been shown to be responsible, in part, for the ability of mmLDL to activate the endothelium to bind monocytes *(13)*. Aldehydes similar to POVPC, such as 1-palmitoyl-(1-steraroyl) 2-(9-oxononanoyl)-,1-palmitoyl-(1-stearoyl)-2-(8-oxooctanoyl)-phosphatidylcholine are also generated in oxLDL, along with aldehydes such as 9-oxononanoyl, 8-oxooctanoyl, and 5-oxovaleroyl esters of cholesterol *(14,15)*. Phosphatidylcholine, a major constituent of LDL and cell membranes of endothelial, smooth, skeletal and cardiac muscles is usually esterified with an arachidonic acid at the *sn-2* position. Thus, POVPC and its analogs are likely to be the major aldehydes generated from the oxidation of lipoproteins and membrane lipids of most cells.

Lipid-Derived Aldehydes and Atherosclerosis

Overwhelming evidence suggests that lipid-derived aldehydes play a key role in atherogenesis. High concentrations of HNE and related aldehydes such as malondialdehyde (MDA) and acrolein are generated during the oxidation of LDL *(2,4,16)*. Once formed, these aldehydes form covalent adducts with nucleophilic amino acid side chains of apoB protein in the LDL particle, thus neutralizing its surface charge and triggering its uptake by the scavenger receptor located on macrophages and SMC *(17,18)*. Antibodies raised against the protein adducts of these aldehydes stain atherosclerotic lesions *(17-20)*, and high titers of auto-antibodies of protein-aldehyde adducts are present in sera of humans with peripheral or coronary artery disease *(21)*, and in apoE-null *(22)* or LDLR-deficient mice *(23)*. Significantly, monoclonal antibodies raised against oxLDL *(24)* or autoantibodies cloned from apoE-null mice *(25)* specifically recognize epitopes generated in oxidized phospholipid or phospholipid-protein adducts derived from phosphatidyl choline, particularly POVPC. Several such antibodies also recognize protein-HNE adducts. In addition to atherosclerotic lesions, protein-HNE antibodies also stain focal areas of neointima after balloon injury *(26)* or during vasculitis *(27)*, indicating increased formation of lipid-derived aldehydes during inflammation and restenosis. Interestingly, antibodies recognizing POVPC also bind to apoptotic cells *(28)*, indicating widespread formation of phospholipid aldehydes during oxidative stress, which is a key component of atherosclerosis.

Increased generation of aldehydes during atherosclerosis suggests that they contribute to disease progression. As mentioned, POVPC activates endothelial cells to bind monocytes *(13)*. In addition, lipid aldehydes may mediate some of the toxic effects of oxLDL on the endothelium. Several injurious effects of oxLDL on endothelial cells have been described including apoptosis *(29,30)*.

Toxicity due to oxLDL has also been suggested to be the underlying reason for the morphological changes and focal defects in the integrity of the vascular endothelium of atherosclerotic arteries, which is an important determinant of plaque stability *(31)*.

Upregulation of ICAM-1 by Aldehydes and Adhesion of Monocytic Cells to Endothelial Cells

The endothelium is a critical regulator of lesion formation and progression. The primary initiating event in lesion formation in atherosclerotic mice is the accumulation of LDL in the subendothelial matrix in proportion to the circulating levels of LDL, which diffuses passively through the endothelial junctions and is retained in the vessel wall by the interactions between the apoB of the LDL particle and the matrix proteoglycans *(1)*. In the intima, LDL and related apoB containing proteins i.e., lipoprotein(a), a lipid rich LDL-like particle containing an additional poly peptide-apolipoprotein(a) molecule attached to apolipoprotein B, undergo a series of modifications including oxidation, lipolysis, proteolysis and aggregation, of which lipid peroxidation is believed to be the key transformation that renders the intimal LDL atherogenic. The initial oxidation of LDL gives rise to the minimally modified LDL (mmLDL) that has proinflammatory properties but is not recognized by the scavenger receptor. The proinflammatory properties of mmLDL contained primarily in its oxidized phospholipids, cause an increased generation of adhesion molecules and growth factors.

We studied the effect of lipid peroxidation-derived aldehydes HNE and acrolein, on the induction of ICAM-1 on endothelial cells (EC). Human umbilical vein endothelial cells (HUVEC) were grown to subconfluency in 12-well dishes. The EC medium containing growth factors and serum was removed. Cells were rinsed once in DMEM-F12 medium and replaced with 10 ml of DMEM-F12. Acrolein (1-5 µM) and HNE (5-10 µM) were incubated with EC for 1h at 37 °C. Following aldehyde exposure, the medium was removed and replaced with the complete growth medium. Cells were stimulated with TNF-α (10 ng/ml) for 24h. The viability of EC appeared normal after treatment. The typical adherent morphology of EC was maintained. Cells were harvested, lysed, and analyzed on SDS gels followed by western blotting. By itself, neither acrolein nor HNE had any effect on the expression of ICAM-1. However, in the presence of TNF-α, ICAM-1 expression was augmented > 3-fold in the presence of 5 µM acrolein and HNE (*Figure 2*). Upregulation of adhesion molecules in the endothelium will result in adhesive interactions of neutrophils and monocytes with the vessel wall. We, therefore, further studied the effect of these aldehydes on the adhesiveness of monocytic cells on the endothelial cells. HUVEC were grown to confluency in 24 well dishes in DMEM medium containing 10% fetal calf serum. Prior to the experiment, the cells were washed with serum-free DMEM medium. The cells were incubated without or with 5μM HNE in serum-free DMEM medium at 37 °C for 1 h. The incubation medium was aspirated out, and the cells were washed with serum-free DMEM medium (to remove any adherent HNE). The cells were then stimulated with TNF-α (10 ng/ml) for 24 h in DMEM medium containing 10% fetal calf serum. Viability of the cells appeared normal after treatment. The cells were washed with RPMI medium containing 5% fetal calf serum. The U937 cells were cultured in RPMI medium containing 7.5% fetal calf serum. Before the experiment, the cells were washed with Hank's balanced salt solution (HBSS) containing 0.1% BSA, counted and resuspended in HBSS containing 0.1% BSA, 1 mM $CaCl_2$ and 1 mM $MgCl_2$ at a concentration of 10^6 cells/ml. Aliquots of these cells (25-100 x 10^3 cells) were then gently layered on HUVEC and allowed to adhere for 30 min at 37°C. The incubation medium was aspirated out, and the cells were washed with HBSS containing 0.1% BSA. Adherence of the U937 cells on HUVEC was examined under the microscope. Under these experimental conditions, more than 50% U937 cells adhered to HUVEC. The cells bound to HUVEC were then treated with 0.25% Rose-Bengal in PBS (100 μl/well) for 5 min at room temperature. The stain was aspirated and the cells were washed with PBS. Aliquots of 200 μl of a solution of ethanol:PBS (1:1) were added to each well and allowed to stand for 30 min at room temperature to release the stain. Optical density of the released stain was measured at 570 nm. HUVEC without the addition of U937 cells were used to estimate non-specific dye binding.As shown in Figure 3, EC treated with HNE or acrolein had significantly higher adhesiveness for the monocytic cells (U937 cells). An increase in the number of U937 cells resulted in an appreciable increase in their adhesiveness to aldehyde-treated EC. This is likely due to the increased availability of the integrin receptors on U937 cells, mediating enhanced interaction with EC through ICAM-1. Increased adhesion of monocytic cells to endothelial cells could lead to the development and progression of atherosclerotic lesions.

Metabolism of HNE in Human Endothelial Cells

Using HNE as a model lipid-derived aldehyde, we examined the aldehyde-metabolizing processes in vascular endothelial cells. Human aortic endothelial cells (3-4 x 10^6) were incubated with 5 μM [^{3}H]-HNE in Ringer-HEPES for 30 min at 37°C. Approximately 95% of HNE was metabolized under these conditions. After 30 min of incubation, the total radioactivity recovered in the incubation medium was 78 ± 8% and < 1% radioactivity was covalently bound to cell constituents. The metabolites extruded in the incubation medium were separated on HPLC as described before (32). The individual peaks were assigned

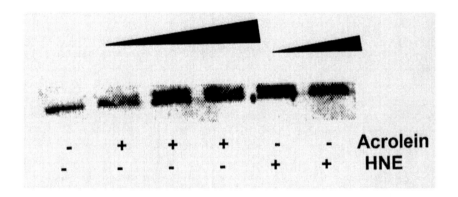

Figure 2. Induction of the expression of ICAM-1 by acrolein and HNE in TNF-α stimulated endothelial cells.

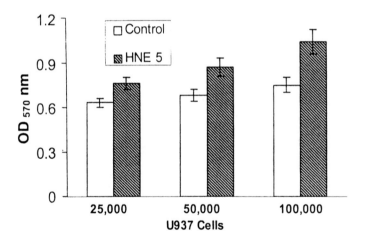

Figure 3. HNE promotes the adhesion of U937 cells to endothelial cells.

to specific HNE metabolites on the basis of the retention time of the synthesized standards (Figure 4). Since the retention time of Peak I corresponded to the glutathione conjugates of HNE and DHN, the peak was collected and analyzed by electrospray mass spectrometry (ESI-MS). This showed a predominant peak with m/z 466.2, which was assigned to GS-DHN (Figure 5). A minor peak with m/z 464.2 due to GS-HNE was also evident, along with a small peak at 446.1. On the basis of the ESI-MS spectra of synthetic GS-HNE, this peak was assigned to the dehydration product of GS-HNE. However, total GS-HNE metabolites in both these peaks was < 30% of the GS-DHN signal, indicating that the extruded conjugate is predominantly GS-DHN. The other major peak (peak V) accounted for 23% of the total radioactivity eluted from the column and this peak had a retention time identical to synthetic HNA. The identity of this peak as HNA was established by gas chromatography-mass spectroscopy (data not shown).

In addition to endothelial cells, we also studied the metabolism of HNE in other cardiovascular tissues. Our observations show that similar to endothelial cells, oxidation to carboxylic acid, glutathiolation and subsequent reduction of the glutathionyl-aldehyde conjugate are the major biochemical pathways for the metabolism of HNE in the cardiovascular tissues (Table I) *(32-36)*.

Role of Aldose Reductase in Aldehyde Metabolism

Our previous studies have shown that aldose reductase (AR) efficiently catalyzes the reduction of medium and long chain 'free' lipid-derived aldehydes (Table II) *(37)*. In addition, AR also catalyzes the reduction of the glutathionyl conjugates of unsaturated aldehydes such as HNE and acrolein. Interestingly, in case of short-chain aldehydes, such as acrolein, conjugation with glutathione leads to a 100-fold increase in the affinity of the enzyme. The reduction of glutathionyl conjugates by AR may be useful in minimizing the reactivity of the aldehyde function, unquenched by glutathiolation.

To test the involvement of AR in the metabolism of HNE in EC, the cells were incubated with sorbinil, the AR inhibitor. After pre-incubation for 30 min with 100 µM sorbinil, 5.0 µM [^{3}H]-HNE was added to medium and the incubation was carried out for an additional 30 min, after which the metabolites were separated on HPLC. The ESI-mass spectrum of peak I obtained from cells treated with sorbinil showed a prominent peak with m/z 466.2 corresponding to GS-DHN (Figure 6). However, the peaks at m/z 464.2 and 446.1due to GS-HNE and its dehydration product were much more prominent in the presence of sorbinil compared to the untreated cells (Figure 5).

The role of AR in the metabolism of lipid-derived aldehydes is further supported by the observation that exposure of vascular smooth muscle cells to HNE or hydrogen peroxide leads to an increase in AR mRNA and protein, and that inhibition of AR accentuates HNE toxicity *(34,38)*. An antioxidant role of

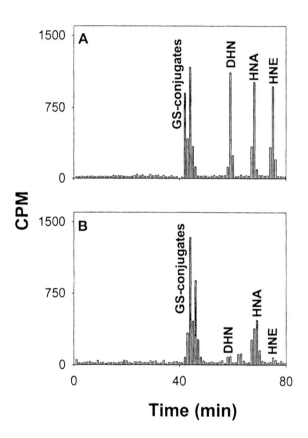

Figure 4. HPLC analysis of HNE metabolites. A) Chromatography for the separation of synthetic GS-HNE/GSDHN, DHN, HNA and HNE. B) The elution profile of radioactivity recovered from the medium in which human aortic endothelial cells were incubated with [³H]-HNE for 30 min.

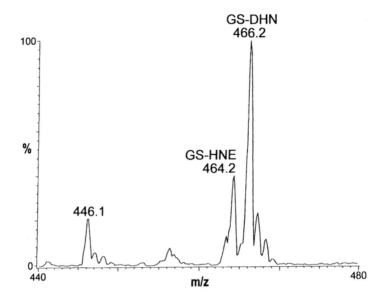

Figure 5. ESI-MS of the incubation medium of the endothelial cells incubated with HNE.

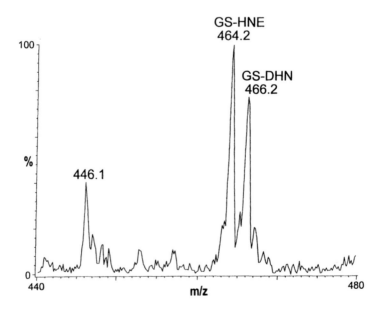

Figure 6. Inhibition of the reduction of GS-HNE by sorbinil in endothelial cells.

Table I. Cardiovascular Metabolism of [^{3}H]-HNE

	GS- HNE/ GS-DHN	Metabolites (%) DHN	HNA	Others
Rat				
Heart Perfusate	33	3	61	3
Vascular Smooth Muscle Cells	60	4	27	9
Red Blood Cells	28	3	27	2
Human				
Red Blood Cells	28	4	42	26
Vascular Smooth Muscle Cells	40	3	46	11
Aortic Endothelial Cells	65	5	23	7

Abbreviations: Glutathionyl 4-hydroxy *trans*-2-nonanal (GS-HNE), glutathionyl dihydoxynonene (GS-DHN), dihydoxynonene (DHN) and 4-hydroxy nonanoic acid (HNA)

Table II. Steady-State Kinetic Parameters of Recombinant Human Heart Aldose Reductase with Aldehydes

Substrate	$K_m(mM)$	$k_{cat}(min^{-1})$
propanal	4.89±0.41	27.0
butanal	0.058 ± 0.005	25.4
hexanal	0.007 ± 0.001	25.6
nonanal	0.026 ± 0.004	30.1
dodecanal	0.041 ± 0.01	36.2
hexadecanal	0.052 ± 0.01	34.7
HNE	0.03 ± 0.009	32.1
GS-HNE	0.02 ± 0.008	29.7
acrolein	0.80 ± 0.21	37.6
GS-propanal	0.007 ± 0.001	36.3

Abbreviations: Glutathionyl 4-hydroxy *trans*-2-nonanal (GS-HNE), glutathionyl propanal (GS-propanal)

AR has also been proposed by Rittner *et al.*, *(27)*, who reported that during vasculitis, the expression of AR is specifically enhanced in areas of high HNE formation. In this model, AR inhibitors increased the concentration of free HNE and protein-HNE adducts, accompanied by a 3-fold increase in the number of apoptotic cells *(27)*. In agreement with previous observations *(39)*, they reported little or no AR in the quiescent, normal SMC of the vessel wall and specific upregulation of the enzyme during oxidative stress. As compared to SMC, the vascular endothelium is one of the most AR-rich tissues and high levels of AR are constitutively expressed in these cells *(39)*. High expression of AR in the endothelial cells would facilitate efficient removal of reactive aldehydes to which the endothelial cells are constantly being exposed via blood-borne toxicants and pollutants. Since oxLDL is the richest *in vivo* source of lipid-derived aldehydes, it follows that removal of these aldehydes by AR is likely to be the key determinant of the endothelial responses to oxLDL.

In contrast to other pathways of metabolism, AR-catalyzed reduction diminishes the reactivity of the aldehyde and may represent true detoxification. Although, HNE readily forms glutathione conjugates *(4)*, the formation of GS-HNE may not, by itself, be sufficient for detoxification. The GS-acrolein conjugate is markedly nephrotoxic *(40)*, the GS-conjugates of hexenal and nonadienal induce DNA damage *(41)* and GS-acrolein is a more potent stimulator of oxygen radical formation than acrolein *(42)*. Therefore, reduction of the glutathione-aldehyde conjugates by AR may be necessary to substantially annul the reactivity of the conjugate and to diminish their participation in free radical generation. Reduction could also facilitate extrusion of the conjugate from the cell.

AR Catalyzes the Reduction of the Phospholipid 'Core' Aldehyde – POVPC

To test the efficiency of AR in catalyzing the reduction of the phospholipid core aldehydes, we used POVPC as a representative core aldehyde. POVPC was a generous gift from Dr. A. D. Watson (University of California, Los Angeles), and was suspended and stored at -20 °C in chloroform. An aliquot of the solution was removed, evaporated under N_2 and dissolved in methanol:water:acetic acid (50:50:0.1 v/v/v) and injected into ESI-MS. The aldehyde formed a well-resolved ion with a m/z value of 594.4 (Fig. 7A). When incubated with 0.1 unit of recombinant AR and 0.1 mM NADPH in 0.15 M potassium phosphate, pH 7.0, the aldehyde was reduced with a kinetic efficiency equal to that of HNE. After 30 min of incubation, the enzyme protein was removed by ultrafiltration and the reaction mixture was desalted using a pipette tip packed with C_{18} resin (Zip-Tip; Millipore). The desalted sample was suspended in methanol:water:acetic acid solution and analyzed by ESI/MS. As

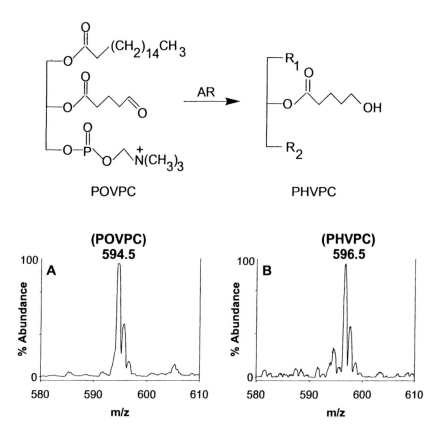

Figure 7. AR catalyzes the reduction of POVPC.

shown in Figure 7B, incubation with AR led to complete reduction of the aldehyde as evidenced by an increase in its m/z ratio by 2. These data provide key evidence that AR reduces the core aldehyde, POVPC.

Reduction may also be important for inactivating POVPC. Chemical reduction of POVPC by sodium borohydride abolishes its ability to activate endothelial cells to bind monocytes (13), whereas, other transformations of POVPC, such as oxidation or hydrolysis, are likely to be less effective. Due to low electrophilicity, POVPC is unlikely to be conjugated to glutathione. Upon oxidation, POVPC forms 1-palmitoyl-2-glutaryl-3-phosphocholine (PGPC), which is as potent as POVPC in activating the endothelial cells (13). In addition, even though PAF-AH mediated hydrolysis has been suggested to protect against free radical injury (43), it leads to the generation of lyso-PC and a saturated semialdehyde. Both these products are highly reactive. LysoPC alters membrane conformation and inhibits cell metabolism (17), whereas, semialdehydes disrupt a host of cellular functions, as exemplified by the extensive tissue degeneration and metabolic disorders in patients with inherited succinate semialdehdye dehydrogenase deficiency (44). Thus, the protective effects of PAF-AH may depend upon subsequent metabolism by enzymes such as AR, and even though multiple pathways of aldehyde metabolism exist, AR-catalysis is likely to be the most significant mode of aldehyde detoxification. In addition, AR catalysis may have other protective effects. Reduction of medium to long-chain aldehydes by AR leads to the formation of relatively inert alcohols. These alcohols are principal precursors of plasmalogens, which do not propagate lipid peroxidation reactions (45), and, therefore, decrease the sensitivity of biological membrane to oxidative stress. Thus, by generating metabolically inert alcohols, AR catalysis will not only inactivate phospholipid aldehydes, but also facilitate phospholipid remodeling so as to enhance cellular resistance to subsequent oxidative stress.

Acknowledgements

This study was supported in part by NIH grants HL65618, HL 59378, and HL 55477.

References

1.   Lusis, A. J. *Nature* **2000**, *407*, 233-241.
2.   Witztum, J. L.; Steinberg, D. *J. Clin. Invest.* **1991**, *88*, 1785-1792.
3.   Chan, H. W-S. *Autoxidation of Unsaturated Lipids*; London, Academic Press, 1987.

4. Esterbauer, H.; Schaur, R. J.; Zollner, H. *Free Radic. Biol. Med.* **1991**, *11*, 81-128.

5. Benedetti, A.; Comporti, M.; Esterbauer, H. *Biochem. Biophys. Acta.* **1980**, *620*, 281-296.

6. Kruman, I.; Bruce-Keller, A. J.; Bredesen, D.; Waeg, G.; Mattson, M. P. *J. Neurosci.* **1997**, *17*, 5089-5100.

7. Bhatnagar, A. *Circ. Res.* **1995**, *76*, 293-304.

8. Kristal, B. S.; Park, B. K.; Yu, B. P. *J. Biol. Chem.* **1996**, *271*, 6033-6038.

9. Page, S.; Fischer, C.; Baumgartner, B.; Haas, M.; Kreusel, U.; Loidl, G.; Hayn, M.; Ziegler-Heitbrock, H. W.; Neumeier, D.; Brand, K. *J. Biol. Chem.* **1999**, *274*, 11611-11618.

10. Ji, C.; Kozak, K. R.; Marnett, L. J. *J. Biol. Chem.* **2001**, *276*, 18223-18228.

11. Okada, K.; Wangpoengtrakul, C.; Osawa, T.; Toyokuni, S.; Tanaka, K.; Uchida, K. *J. Biol. Chem.* **1999**, *274*, 23787-23793.

12. Stremler, K. E.; Stafforini, D. M.; Prescott, S. M.; Zimmerman, G. A.; McIntyre; T. M. *J. Biol. Chem.* **1989**, *264*, 5331-5334.

13. Watson, A. D.; Leitinger, N.; Navab, M.; Faull, K. F.; Horkko, S.; Witztum, J. L.; Palinski, W.; Schwenke, D.; Salomon, R. G.; Sha, W.; Subbanagounder, G.; Fogelman, A.M.; Berliner, J.A. *J. Biol. Chem.* **1997**, *272*, 13597-13607.

14. Kamido, H.; Kuksis, A.; Marai, L.; Myher, J. J. *J. Lipid Re.s* **1995**, *36*, 1876-1886.

15. Karten, B.; Boechzelt, H.; Abuja, P. M.; Mittelbach, M.; Sattler, W. *J. Lipid Res.* **1999**, *40*, 1240-1253.

16. Steinberg, D. *J. Biol. Chem.* **1997**, *272*, 20963-20966.

17. Bolgar, M. S.; Yang, C. Y.; Gaskell, S. J. *J. Biol. Chem.* **1996**, *271*, 27999-28001.

18. Uchida, K.; Kanematsu, M.; Morimitsu, Y.; Osawa, T.; Noguchi, N.; Niki, E. *J. Biol. Chem.* **1998**, *273*, 16058-16066.

19. Yla-Herttuala, S.; Palinski, W.; Rosenfeld, M. E.; Parthasarathy, S.; Carew, T. E.; Butler, S.; Witztum, J. L.; Steinberg, D. *J. Clin. Invest.* **1989**, *84*, 1086-1095.

21. Salonen, J. T.; Yla-Herttuala, S.; Yamamoto, R.; Butler, S.; Korpela, H.; Salonen, R.; Nyyssonen, K.; Palinski, W.; Witztum, J. L. *Lancet* **1992**, *339*, 883-887.

22. Palinski, W.; Horkko, S.; Miller, E.; Steinbrecher, U. P.; Powell, H. C.; Curtiss, L. K.; Witztum, J. L. *J. Clin. Invest.* **1996**, *98*, 800-814.

23. Tsimikas, S.; Palinski, W.; Witztum, J. L. *Arterioscler. Thromb. Vasc. Biol.* **2001**, *21*, 95-100.

24. Itabe, H.; Takeshima, E.; Iwasaki, H.; Kimura, J.; Yoshida, Y.; Imanaka, T.; Takano, T. *J. Biol. Chem.* **1994**, *269*, 15274-15279.

25. Horkko, S.; Bird, D. A.; Miller, E.; Itabe, H.; Leitinger, N.; Subbanagounder, G.; Berliner, J. A.; Friedman, P.; Dennis, E. A.; Curtiss, L. K.; Palinski, W.; Witztum, J. L. *J. Clin. Invest.* **1999**, *103*, 117-128.

26. Ruef, J.; Hu, Z. Y.; Yin, L. Y.; Wu, Y.; Hanson, S. R.; Kelly, A. B.; Harker, L. A.; Rao, G. N.; Runge, M. S.; Patterson, C. *Circ. Res.* **1997**, *81*, 24-33.

27. Rittner, H. L.; Hafner, V.; Klimiuk, P. A.; Szweda, L. I.; Goronzy, J. J.; Weyand, C.M. *J. Clin. Invest.* **1999**, *103*, 1007-1013.

28. Chang, M. K.; Bergmark, C.; Laurila, A.; Horkko, S.; Han, K. H.; Friedman, P.; Dennis, E. A.; Witztum, J. L. *Proc. Natl. Acad. Sci. U. S. A.* **1999**, *96*, 6353-6358.

29. Escargueil-Blanc, I.; Meilhac, O.; Pieraggi, M. T.; Arnal, J. F.; Salvayre, R.; Negre-Salvayre, A. *Arterioscler. Thromb. Vasc. Biol.* **1997**, *17*, 331-339.

30. Harada-Shiba, M.; Kinoshita, M.; Kamido, H.; Shimokado, K. *J. Biol. Chem.* **1998**, *273*, 9681-9687.

31. Bjorkerud, B.; Bjorkerud, S. *Arterioscler. Thromb. Vasc. Bio.* **1996**, *16*, 416-424.

32. Srivastava, S.; Conklin, D. J.; Liu, S-Q; Prakash, N.; Boor, P. J.; Srivastava, S. K.; Bhatnagar, A. *Atherosclerosis* **2001**, *158*, 339-350.

33. Srivastava, S.; Chandra, A.; Wang, L. F.; Seifert, W. E. J.; DaGue, B. B.; Ansari, N. H.; Srivastava, S.K.; Bhatnagar, A. *J. Biol. Chem.* **1998**, *273*, 10893-10900.

34. Ruef, J.; Liu, S. Q.; Bode, C.; Tocchi, M.; Srivastava, S.; Runge, M. S.; Bhatnagar, A. *Arterioscler. Thromb. Vasc. Biol.* **2000**, 20, 1745-1752.

35. Srivastava, S.; Dixit, B. L.; Cai, J.; Sharma, S.; Hurst, H. E.; Bhatnagar, A.; Srivastava, S. K. *Free Radic. Biol. Med.* **2000**, *29*, 642-651.

36. Srivastava, S.; Liu, S-Q.; Conklin, D. J.; Zacarias, A.; Srivastava, S. K.; Bhatnagar, A. *Chem. Biol. Interact.* **2001**, *30-32*, 563-571.

37. Srivastava, S.; Watowich, S. J.; Petrash, J. M.; Srivastava, S. K.; Bhatnagar, A. *Biochemistry* **1999**, *38*, 42-54.

38. Spycher, S. E.; Tabataba-Vakili, S.; O'Donnell, V. B.; Palomba, L.; Azzi, A. *FASEB. J.* **1997**, *11*, 181-188.

39. Ludvigson, M. A.; Sorenson, R. L. *Diabetes* **1980**, *29*, 438-449.

40. Horvath, J. J.; Witmer, C. M.; Witz, G. *Toxicol. Appl. Pharmacol.* **1992**, *117*, 200-207.

41. Dittberner, U.; Eisenbrand, G.; Zankl, H. *Mutat. Res.* **1995**, *335*, 259-265.

42. Adams, J. D. J.; Klaidman, L. K. *Free Radic. Biol. Med.* **1993**, *15*, 187-193.

43. Matsuzawa, A.; Hattori, K.; Aoki, J.; Arai, H.; Inoue, K. *J. Biol. Chem.* **1997**, *272*, 32315-32320.

44. Roe, C. R.; Struys, E.; Kok, R. M.; Roe, D. S.; Harris, R. A.; Jakobs, C. *Mol. Genetic. Metab.* **1998**, *65*, 35-43.

45. Reiss, D.; Beyer, K.; Engelmann, B. *Biochem. J.* **1997**, *323*, 807-814.

Aldo-Keto Reductases and Exogenous Toxicants

Tobacco-Related Carcinogens

Chapter 5

Competing Roles of Reductases in the Detoxification of the Tobacco-Specific Nitrosamine Ketone NNK

Edmund Maser[1] and Ursula Breyer-Pfaff[2]

[1]Department of Experimental Toxicology, School of Medicine, University of Kiel, 24105 Kiel, Germany
[2]Department of Pharmacology and Toxicology, Section on Toxicology, University of Tübingen, 72074 Tübingen, Germany

Smoking is linked to lung cancer, yet only a small fraction of smokers develop this disease. Identification of genetic, environmental and nutritional factors that affect lung-cancer risk might explain why some smokers are more likely to develop lung cancer than others. The balance between metabolic activation and detoxification is critical in determining the susceptibility to lung cancer upon exposure to the tobacco-specific nitrosamine 4-(N-methyl-N-nitrosamino)-1-(3-pyridyl)-1-butanone (NNK). In man, activation of NNK occurs by cytochrome P-450 (CYP)-mediated monooxygenation, whereas detoxification is initiated by carbonyl reduction of NNK to its corresponding alcohol 4-(N-methyl-N-nitrosamino)-1-(3-pyridyl)-1-butanol (NNAL) which is glucuronosylated and excreted. Because carbonyl reduction of NNK is essential for glucuronosylation, producing the detoxified metabolite NNAL-Gluc, the equilibrium between NNK and NNAL is suspected to play a key role in the carcinogenicity of NNK and its organospecificity. Our studies are focused on the identification, expression and activity of NNK carbonyl reductases. Five different NNK carbonyl reductases have been identified in man: Two are members of the short-chain dehydrogenases/reductases, namely 11β-hydroxysteroid dehydrogenase type 1 and cytosolic carbonyl reductase and three are members of the aldo-keto reductase (AKR) superfamily, namely AKR1C1, AKR1C2, and AKR1C4. All enzymes have been purified to homogeneity from human liver and characterized with respect to NNK carbonyl reduction. Since (S)-NNAL is a more potent murine tumorigen than (R)-NNAL, the enantioselectivity of the purified human enzymes was determined. We propose that the extent of expression and activity of NNK reductases strongly influences the tissue selectivity of and interindividual susceptibility to NNK-mediated cancer.

A worldwide public health problem is the increasing number of deaths due to lung cancer, especially in the developing countries. About 90% of lung cancers in men and 79% in women are directly attributable to smoking (reviewed in (1)). Compared with non-smokers, smokers have a 10-fold greater risk of dying from lung cancer, and in heavy smokers this risk increases to 15- to 25-fold (2). Although at least 55 carcinogens have been identified in cigarette smoke, there is strong evidence that tobacco-specific nitrosamines play an important role in cancer induction by tobacco products, among which 4-(N-methyl-N-nitrosamino)-1-(3-pyridyl)-1-butanone (NNK) is the most carcinogenic (reviewed in (3)). NNK is a potent and selective inducer of adenocarcinoma of the lung in rodents. In F344 rats, lung tumors were induced regardless of the route of administration, including in drinking water, or by subcutaneous injection, gavage, oral swabbing or intravesicular injection (reviewed in (3)). Uptake of NNK by humans has been conclusively demonstrated by quantification of its metabolites in urine of smokers (4-8).

NNK is a Procarcinogen That Requires Metabolic Activation

There are competing metabolic pathways for the activation and detoxification of NNK. NNK is metabolically activated by α-hydroxylation at the carbons adjacent to the N-nitroso group, catalyzed by various cytochrome P-450 (CYP) enzymes (reviewed in (3)). Of the human CYPs examined, CYP1A1, CYP1A2, CYP2A6, CYP2B6, CYP2D6, CYP2E1, CYP2F1 and CYP3A4 catalyze the formation of carcinogenic NNK metabolites (reviewed in (9)).

Involvement of other enzymes, such as lipoxygenases and cyclooxygenases in oxidative metabolism of NNK has also been reported (*10*). Metabolic activation occurs via both methylene hydroxylation to produce a methylating species and methyl hydroxylation to produce a pyridyloxobutylating species (reviewed in (*3*)). The resulting electrophilic intermediates can alkylate DNA bases, thereby inducing *p53* tumor-suppressor gene inactivation and *K-ras* proto-oncogene activation. The initial event in NNK-induced lung tumorigenesis is the formation of N7- and O6-methylguanine (O6mG) or pyridyloxobutylated DNA. Significant amounts of these DNA lesions are found in lung cancers and are regarded to be responsible for malignant mutations of normal tissue. For example, studies of lung carcinogenesis induced by NNK in A/J mice and rats have shown a strong correlation between lung tumor formation and levels of O6mG adducts. O6mG is a major promutagenic adduct which leads to GC→ AT transitional mispairing and an activation of the *K-ras* proto-oncogene in the A/J mouse lung. Other studies in A/J mice focused on NNK-induced pyridyloxobutylated DNA and found that this bulky DNA adduct inhibits repair of O6mG (reviewed in (*3*)).

Carbonyl Reduction Initiates Detoxification of NNK

A major pathway of NNK metabolism is its conversion to its carbonyl reduction product 4-(*N*-methyl-*N*-nitrosamino)-1-(3-pyridyl)-1-butanol (NNAL), which has been evaluated in cultured human tissues and human liver and lung microsomes (*11,12*). NNAL-*N*-oxide is another NNK detoxification product found in human urine, although its level is relatively low (*6*), indicating that pyridine *N*-oxidation is a relatively minor detoxification pathway in humans compared to carbonyl reduction/glucuronidation.

Despite the fact that NNAL is carcinogenic itself in rats and mice (*3*), carbonyl reduction of NNK to NNAL can be considered as a detoxification pathway since NNAL is rapidly glucuronosylated and excreted. In other words, carbonyl reducing enzymes initiate the final detoxification of NNK by reducing the ketone moiety of NNK to provide the hydroxy function in NNAL necessary for glucuronosylation (*13*). Both NNAL and NNAL-Gluc were detected in the urine of smokers, smokeless tobacco users, and nonsmokers exposed to environmental tobacco smoke (*4-8*).

It has been suggested that the NNK/NNAL ratio in a given tissue is extremely important for organ susceptibility to the development of cancer (*13-16*). Thus, the high level of NNAL formed by rat liver facilitates elimination of NNK, consequently resulting in low tumor yield in this organ, while the lack of NNAL formation in rat nasal mucosa leads to a high tumor incidence (*16*).

The NNAL Enantiomers Differ in Their Carcinogenicity

NNAL is a chiral compound while NNK is prochiral. Recent studies indicate that stereochemical aspects of NNAL and NNAL-Gluc formation could play an important role in the metabolic detoxification of NNAL as well as its carcinogenic activity (*17*). Bioassay results in A/J mice indicate that the carcinogenic potency of (*S*)-NNAL was equal to that of the parent compound NNK, while (*R*)-NNAL was less potent (*17*). It was hypothesized that this difference could be caused by the difference in the *in vivo* disposition profile of the two enantiomers rather than a difference in their intrinsic tumorigenic activity. *In vivo*, (*R*)-NNAL is more rapidly eliminated due to (*R*)-NNAL-Gluc formation and biliary excretion, while (*S*)-NNAL is subject to α-hydroxylation as well as stereoselective localization in the lung. The latter may contribute to lung selectivity of NNK carcinogenesis (reviewed in (*18*)).

Individual Differences in Lung Cancer Susceptibility

Only a small portion of all habitual smokers develop lung cancer, suggesting the existence of susceptibility genes. There are wide variations among individuals in their ability to metabolize NNK, and these differences may be linked to lung cancer susceptibility. While the involvement of several CYP enzymes in the activation of NNK is well documented (*3*), the relationship between lung cancer risk, tobacco smoke and CYP activities has been controversial (*19,20*). On the other hand, NNK reduction to NNAL was estimated to range between 39% and 100% of the NNK dose in smokers (*4*). Moreover, the quantification of metabolites in smokers' urine revealed that NNK is predominantly metabolized by carbonyl reduction in humans (*4,5,7,8,21,22*). Thus, it appears that the NNK/NNAL ratio in an individual is determined by the activity of NNK carbonyl reductases, and this may determine carcinogenic outcome

The Emerging Role of NNK Reductases in the Protection Against Tobacco-Smoke Derived Lung Cancer

Since the enzyme systems responsible for NNAL formation have not been well characterized, we focused our investigation on the identification, purification and characterization of NNK carbonyl reductases. Five different enzymes catalyzing NNK carbonyl reduction in man have been identified in our studies (Figure 1).

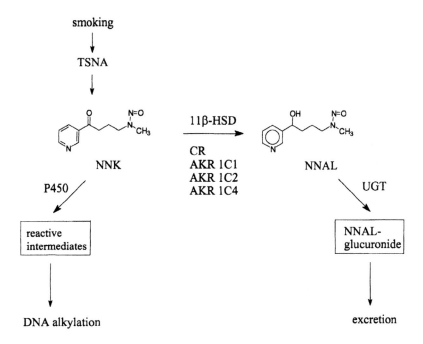

Figure 1. Simplified scheme of the metabolic fate of the tobacco-derived carcinogen NNK. The NNK/NNAL equilibrium is determined by the level of expression and activity of NNK carbonyl reductases which initiate NNK detoxification by providing the hydroxyl moiety necessary for glucuronosylation and final excretion. Note that back conversion of NNAL to NNK and pyridine N-oxidation reactions, the latter being of no significance in humans, are not shown. Abbreviations: TSNA, tobacco-specific nitrosamines; CYP, cytochromes P450; UGT, uridine diphosphate glucuronosyl transferase.

These are microsomal 11β-hydroxysteroid dehydrogenase type 1 (11β-HSD 1) (EC 1.1.1.146) *(13,15)* and cytosolic carbonyl reductase (EC 1.1.1.184) *(23)*, where both enzymes belong to the short-chain dehydrogenase/reductase (SDR) superfamily. In addition, three members of the aldo-keto reductase (AKR) superfamily have been shown to mediate NNK carbonyl reduction, namely AKR1C1, AKR1C2 and AKR1C4 (EC 1.3.1.20) *(23)*, previously designated as dihydrodiol dehydrogenases DD1, DD2 and DD4, respectively.

SDR-Type Reductases Mediating NNK Carbonyl Reduction

The short-chain dehydrogenase/reductase (SDR) family contains more than 1000 proteins and is one of the largest protein superfamilies known to date (24). Even though there is very little sequence identity (15-30%) among the family members, two consensus sequences are strictly conserved. One is the GxxxGxG cosubstrate-binding motif near the N-terminus, and the other is the so-called Ser-Tyr-Lys triad that comprises the active site and is critical for the catalytic activity of the SDR members. Despite the low sequence identities, the three-dimensional structures display a highly similar α/β folding pattern with a central β-sheet, typical for a Rossmann-fold. The SDR enzymes accept a wide variety of substrates like sugars, steroids, retinoids, prostaglandins and xenobiotic aldehydes and ketones (25).

The SDRs have been divided into two large families, the "classical" with 250-odd residues and the "extended" with 350-odd residues (25). Based on patterns of charged residues in the cosubstrate-binding region, these families have recently been classified in seven subfamilies of classical SDRs and three subfamilies of extended SDRs. Three further families are novel entities, denoted "intermediate", "divergent" and "complex", encompassing short-chain alcohol dehydrogenases, enoyl reductases and multifunctional enzymes, respectively. Both 11β-HSD 1 and carbonyl reductase are members of the cP2 subfamily of the classical SDRs (26).

11β-Hydroxysteroid Dehydrogenase type 1 (11β-HSD 1)

11β-Hydroxysteroid dehydrogenase (11β-HSD) physiologically catalyzes the interconversion of receptor-active 11-hydroxy glucocorticoids (cortisol) to their receptor-inactive 11-oxo metabolites (cortisone). By this action, 11β-HSD regulates glucocorticoid access to both the glucocorticoid receptor and mineralocorticoid receptor, thereby playing a key role in glucocorticoid homeostasis (27). Two different isoforms have been described so far, 11β-HSD 1 and 11β-HSD 2, which differ in their biological properties and tissue distribution (28,29).

The type 1 isozyme of 11β-HSD is a low affinity NADP(H) dependent dehydrogenase/oxo-reductase, with an apparent K_m for glucocorticoids in the low micromolar range. The enzyme has an almost ubiquitous tissue distribution with highest expression in the liver. Although the enzyme reaction catalyzed by 11β-HSD 1 in vitro is bidirectional, recent evidence suggest it in vivo to function primarily as reductase, reactivating inert 11-oxo glucocorticoids to the corresponding receptor-active 11β-hydroxy forms (30-33). This activity is likely to be of particular importance in tissues such as the liver, adipose tissue and brain, where it may maintain high intracellular levels of glucocorticoids.

Another important characteristic of 11β-HSD 1 is its ability to reduce a variety of xenobiotic carbonyl compounds as the first step in a consecutive detoxification process (34,35). 11β-HSD 1 has been purified from mouse (36) and human (37) liver and was the first enzyme identified to act as NNK carbonyl reductase (13). The kinetic properties of 11β-HSD 1 regarding NNK metabolism are shown in Table I.

**Table I. Kinetics and enantioselectivity of NNAL formation
by purified enzymes**

Enzyme	V_{max}	K_m	(S)-NNAL
11β-HSD 1	245	12	65
Carbonyl reductase	2,800	7	96
AKR1C1	18	0.2	99
AKR1C2	17	0.3	96
AKR1C4	8	0.8	94

Values are means of three to four individual preparations. V_{max} = nmol NNAL / mg x min; K_m = mM; (S)-NNAL is given as percentage from total NNAL formation.

Carbonyl Reductase

Carbonyl reductase (EC 1.1.1.184) is an NADPH-dependent enzyme that catalyzes the reduction of steroids, prostaglandins and a variety of xenobiotics (38,39). However, high non-physiological K_m values and low turnover numbers in animal and human tissues raised some doubts as to the authentic endogenous steroid and prostaglandin substrates (reviewed in (40)), although the enzyme has been suggested to act as 3α,20β-HSD in animal tissues (39). Wermuth showed that quinones derived from polycyclic aromatic hydrocarbons were much better substrates and concluded that carbonyl reductase is the main NADPH-dependent quinone reductase in human liver, playing a significant role in quinone detoxification in man (41). Previous studies on the inhibition pattern of NNK carbonyl reduction led to the suggestion that carbonyl reductase is the major NNK reductase in human liver and lung cytosol (42), a finding that was corroborated with the purified enzyme (23) (Table I).

AKR-Type Reductases Mediating NNK Carbonyl Reduction

The aldo-keto reductases (AKRs) encompass a large superfamily of NAD(P)(H)-dependent oxidoreductases that act on a wide variety of natural and foreign substrates. They all possess the α/β_8-barrel motif characteristic of triose phosphate isomerase (TIM) and contain approximately 320 amino acids per monomer. They catalyze the reduction of aldehydes and ketones, monosaccharides, ketosteroids and prostaglandins. They also catalyze the oxidation of hydroxysteroids and *trans*-dihydrodiols of polycyclic aromatic hydrocarbons and are found in prokaryotes and eukaryotes including yeast, plants, amphibia and mammals (reviewed in (*43*)). A systematic nomenclature for the AKR superfamily was adopted in 1996 (*44*) and was updated in 2003 (*45*). (www.med.upenn.edu/akr). The superfamily contains 114 proteins that distribute over 14 families (AKR1-AKR14).

The AKR1C Subfamily

The AKR1 family contains steroid-hormone transforming enzymes that are implicated in the inactivation and mechanism of action of steroid sex hormones. Four NADP(H)-dependent enzymes comprising the AKR1C subfamily and exhibiting 3α-HSD activity have been identified in human tissues (*46,47*). The enzymes share more than 83% amino acid sequence identity, and have been shown to exhibit broad substrate specificities for 3α-, 17β- and 20α-hydroxysteroids. According to the above mentioned nomenclature they are termed AKR1C1 – AKR1C4. Previously used names for the enzymes are: AKR1C1 = 3α,20α-HSD or dihydrodiol dehydrogenase type 1; AKR1C2 = 3α-HSD type III, bile acid binding protein or dihydrodiol dehydrogenase type 2; AKR1C3 = 3α-HSD type II, 17β-HSD type V or dihydrodiol dehydrogenase X; AKR1C4 = 3α-HSD type I, chlordecone reductase or dihydrodiol dehydrogenase type 4 (*48,49*). *In vitro* characterization revealed that AKR1C1-AKR1C4 display functional plasticity in that they catalyze 3-, 17- and 20-ketosteroid reductase activity and 3α-, 17β- and 20α-hydroxysteroid oxidase activity to varying degrees (*47*). Hence, the enzymes have the capacity to modulate access of active androgens, estrogens and progestins to their cognate receptors (*50*).

AKR1C1 and AKR1C2

AKR1C1 (3α,20α-HSD) has a high k_{cat}/K_m for 20-ketosteroid reduction and is a candidate for the major human 20α-HSD that inactivates progesterone (*47*).

AKR1C2 (3α-HSD type III) has a k_{cat}/K_m for 3-ketosteroids which is 50-fold less than that of AKR1C4. Its preference to act as a 3-ketosteroid reductase and as a 3α-hydroxysteroid oxidase suggests that it acts as a peripheral 3α-HSD (47,51). Importantly, it can interconvert 5α-dihydrotestosterone and 3α-androstanediol in the prostate and may regulate occupancy of the androgen receptor. Based on their primary structure, AKR1C1 differs from AKR1C2 only by seven amino acids (52). Interestingly, AKR1C1 and AKR1C2 play different roles in the cerebral metabolism of neurosteroids. Neuroactive steroids such as 3α,5α-tetrahydroprogesterone and 3α,5α-tetrahydrodeoxycorticosterone have been shown to be synthesized from progesterone in animal brains. Comparison of kinetic constants for the steroids and their precursors among the four human AKR1C members suggests that AKR1C1 and AKR1C2 mediate catabolism and synthesis, respectively, of the neuroactive steroids in the human brain (52).

AKR1C3

AKR1C3 (3α-HSD type II) does not show significant activities towards the neurosteroids. Due to its k_{cat}/K_m values it favours 17-ketosteroid reduction over 3-ketosteroid reduction and therefore functions as a peripheral 3α(17β)-HSD (53). It reduces Δ^4-androstene-3,17-dione to testosterone and estrone to 17β-estradiol and therefore mediates activation of androgens and estrogens.

AKR1C4

AKR1C4 (3α-HSD type I) is liver specific and because of its high k_{cat}/K_m for dihydrosteroids it likely works in concert with 5α-reductase and 5β-reductase in the hepatic clearance of steroid hormones and bile acid biosynthesis (47). In addition to androgens, it may also control the concentration of circulating neuroactive steroids. AKR1C1, 1C2 and 1C4 have been purified from human liver cytosol (54) and their kinetics regarding NNK carbonyl reduction have been determined (23) (Table I).

Relative Importance of the Individual SDR- and AKR-Type NNK Reductases

Estimates on the relative importance of the individual cytosolic enzymes for total NNK carbonyl reduction in human liver can be based on enzyme efficiencies (Table I) and relative enzyme quantities. On the assumption that losses during purification are comparable for the enzymes, ratios of enzyme quantities in cytosol should be roughly 3 : 3 : 1 : 1.5 for AKR1C1, AKR1C2, AKR1C4 and carbonyl reductase (23,54). Multiplication by the enzyme efficiencies (23) results in relative contributions of roughly 20, 20, 1, and 60% to NNAL formation in cytosol by the four soluble enzymes.

On purification of carbonyl reductase and the three AKR1Cs from human liver cytosol, no further NNK reductases were detected. Moreover, the enantioselectivity of NNK reduction in cytosol is in accordance with that of the isolated enzymes (Table I). In contrast, comparison of enantioselectivities as well as inhibition experiments suggest the presence of at least two other NNK reductases in the microsomal fraction, in addition to 11β-HSD 1 (see below).

Enantioselectivity of the SDR- and AKR-Type NNK Reductases

Because NNK is metabolized to both the (S) and (R) enantiomers of NNAL, two diastereomeric glucuronides of NNAL are formed (3). The conjugating enzymes responsible are members of the UDP-glucuronosyltransferase (UGT) family. Stereochemical aspects of NNAL and NNAL-Gluc formation could either decide whether NNAL is detoxified or whether it is carcinogenic (3,17) Species differences are known to exist in NNAL metabolism. Whereas (R)-NNAL-Gluc is the major diastereomer in rodent urine, (S)-NNAL predominates in patas monkey urine (55). Moreover, experiments with toombak users' urine demonstrated that (S)-NNAL-Gluc was the predominant diastereomer in these samples (56). The results of analyses of urine samples from 30 smokers (7) demonstrated that the enantiomeric ratio of NNAL in urine was 54% (S) and 46% (R), whereas the diastereomeric composition of NNAL-Gluc was 68% (S) and 32% (R). Therefore, (R)-NNAL-Gluc is the major detoxification product of NNK in rodents, whereas (S)-NNAL-Gluc seems to be the major detoxification product of NNK in primates including humans.

In support of this assumption, Ren et al (57) reported that human liver microsomes preferentially catalyze the formation of (S)-NNAL-Gluc from racemic NNAL. They identified UGT2B7, which preferentially produces (S)-NNAL-Gluc and is expressed in human liver, as one of the enzymes involved in this reaction.

Because (S)-NNAL is the major enantiomer formed by the purified NNK reductases (Table I) and (S)-NNAL-Gluc is the major glucuronidated diastereomer observed in the urine of smokers, it is predicted that SDR and AKR enzymes will play important roles in the detoxification of NNAL (and therefore, NNK) in human tissues. Using human liver microsomes NNK is reduced to NNAL to contain 70% (R)-enantiomer (58) in contrast to 35% that results from the reduction catalyzed by purified 11β-HSD 1 (Table I). Therefore, at least one other NNAL-forming enzyme should be active in the endoplasmic reticulum of human liver. This assumption is corroborated by preliminary investigations in which solubilized 11β-HSD 1 could be chromatographically separated from unidentified NNK-reductases with high enantioselectivities either for *(R)*- or for *(S)*-NNAL formation (*58*).

Inhibition of SDR- and AKR-Type NNK Reductases by Licorice

A number of specific inhibitors of NNK reductases are known. However, the inhibitor glycyrrhetinic acid, a constituent of licorice, deserves special comment. Licorice is derived from the rhizomes and roots of *Glycyrrhiza glabra* L. The extract contains up to 10% glycyrrhizic acid, a saponin-like glycoside 50 times sweeter than sugar. Currently, licorice and glycyrrhetinic acid (the aglycone of glycyrrhizic acid) are used in a great deal of food products as aromatizers and sweeteners, as well as in the pharmaceutical industry. Clinical studies have shown that licorice has spasmolytic properties and beneficial influences on the healing of gastric ulcers. However, side effects related to high doses of glycyrrhetinic acid ingestion have been described, including cardiac dysfunction, oedema, hypertension, disturbance in body-electrolyte balance, headaches and body weight increase. These negative effects may result from glycyrrhetinic acid inhibition of a variety of enzymes, including the cortisol metabolizing 11β-HSD isoforms (*59*). However, glycyrrhetinic acid is not only a potent (in nM concentrations) inhibitor of 11β-HSD isoforms (*59*), but has recently been shown to inhibit (in μM concentrations) AKR 1C1, 1C2, 1C3 and 1C4 (*60*). Moreover, glycyrrhetinic acid inhibited the formation of NNAL in incubations with human cervical microsomes (*16*).

Therefore, licorice ingestion may attenuate NNK detoxification by the NNK reductases. Moreover, licorice is used as a tobacco additive to alleviate mucosa irritation upon smoking (*61*). The amounts of glycyrrhetinic acid found in cigarette tobacco are 0.11%, which is consistent with a 10% yield from the 1.3% of licorice added to tobacco (*61*). Finally, glycyrrhetinic acid is a known inducer of CYPs (*62*). The resulting increase in NNK activation via CYPs could act synergistically with the inhibition of NNK carbonyl reductases, thereby further aggravating the toxicological consequences of smoking.

Pluripotency of Hydroxysteroid Dehydrogenases

Physiological Role

Steroid hormones act by binding to receptor proteins in target cells, which leads to transcriptional regulation of different gene products and the desired physiological response. In target tissues, however, hydroxysteroid dehydrogenases (HSDs) interconvert potent steroid hormones to their cognate inactive metabolites (and *vice versa*) and thus regulate the occupancy of steroid hormone receptors. In normal cell physiology HSDs therefore function as important prereceptor regulators of signaling pathways by acting as "molecular switches" between receptor-active and receptor-inactive hormone (*63-66*).

Importance of Carbonyl Reducing Enzymes

Interestingly, several HSDs have been shown to catalyze the carbonyl reduction of non-steroidal aldehydes and ketones. Carbonyl reduction is a significant step in the phase I biotransformation of a great variety of aromatic, alicyclic and aliphatic xenobiotic carbonyl compounds (*67*). Compared with the oxidative CYP system, carbonyl reducing enzymes had, for a long time, received considerably less attention. However, the advancement of carbonyl reductase molecular biology has allowed the identification of several carbonyl reducing enzymes, including pluripotent HSDs that are involved in xenobiotic carbonyl compound detoxification in addition to catalyzing the oxidoreduction of physiologic steroid substrates (*40*). Based on their primary structure these enzymes could be classified into either the SDR or AKR protein superfamilies (as dicussed above). As the HSD/carbonyl reductase project continues to progress, it is anticipated that new members of these groups of enzymes, which play important roles in not only the metabolism of xenobiotics but also the biotransformation of a variety of endogenous steroids, are bound to emerge, like for example hitherto unidentified microsomal NNK reductases.

Evolution of Pluripotency

The existence of enzymes that have arisen independently to have a common activity has been repeatedly observed in different enzyme families. In many cases, such analogous enzymes seem to evolve by recruitment of enzymes acting on different but related substrates, i.e. by minor structural change of a protein that leads to a novel specificity or even a new class of reactions. The need to deal with the deleterious effects of toxic aldehydes, ketones and quinones and a changing endocrine environment may have resulted in the evolution of pluripotent HSDs.

With their pluripotent substrate specificities for steroids and NNK the HSDs in this study add to an expanding list of those enzymes that, on the one hand, are capable of catalyzing the carbonyl reduction of non-steroidal xenobiotics, and that on the other hand exhibit great specificity for their physiological substrates (*40*).

Outlook

Molecular epidemiology has contributed to a growing awareness of the importance of susceptibility factors in modulating risks associated with exposure to environmental carcinogens. Knowledge of the prevalence and distribution of genetic susceptibility factors and the ability to identify susceptible individuals or subgroups will have substantial preventive implication, especially if "at risk" genotypes can be correlated to susceptibility to tobacco smoke-derived lung cancer.

Modulation of responsiveness to tobacco smoke carcinogens by host genetic factors may explain why fewer than 20% of smokers will get lung cancer. With respect to NNK, the search for such host genetic factors is currently focused on genes coding for activating cytochrome P450 enzymes and in particular on detoxifying NNK carbonyl reducing enzymes and members of the glucuronosyl transferase family. The discovery that HSDs initiate the detoxification of the tobacco-specific nitrosamine NNK is of great interest and provides the basis for further research. It will be of interest in the future to correlate the expression of these enzymes in diseased and normal subjects in an effort to understand their protective role against tobacco-smoke related lung cancer.

References

1. Djordjevic, M. V.; Stellman, S. D.; Zang, E. *J. Natl. Cancer Inst.* **2000,** *92,* 106-111.
2. Carbone, D. *Am. J. Med.* **1992,** *93,* 13S-17S.
3. Hecht, S. S. *Chem. Res. Toxicol.* **1998,** *11,* 559-603.
4. Carmella, S. G.; Akerkar, S.; Hecht, S. S. *Cancer Res.* **1993,** *53,* 721-724.
5. Carmella, S. G.; Akerkar, S. A.; Richie, J. P.; Hecht, S. S. *Cancer Epidemiol. Biomarkers & Prev.* **1995,** 4, 635-652.
6. Carmella, S. G.; Borukhova, A.; Akerkar, S. A.; Hecht, S. S. *Cancer Epidemiol. Biomarkers Prev.* **1997,** *6,* 113-120.
7. Carmella, S. G.; Ye, M.; Upadhyaya, P.; Hecht, S. S. *Cancer Res.* **1999,** *59,* 3602-3605.
8. Carmella, S. G.; Le Ka, K. A.; Upadhyaya, P.; Hecht, S. S. *Chem. Res. Toxicol.* **2002,** *15,* 545-550.
9. Su, T.; Bao, Z.; Zhang, Q. Y.; Smith, T. J.; Hong, J. Y.; Ding, X. *Cancer Res.* **2000,** *60,* 5074-5079.
10. Smith, T. J.; Stoner, G. D.; Yang, C. S. *Cancer Res.* **1995,** *55,* 5566-5573.
11. Castonguay, A.; Stoner, G. D.; Schut, H. A.; Hecht, S. S. *Proc. Natl. Acad. Sci. U.S.A.* **1983,** *80,* 6694-6697.

12. Smith, T. J.; Guo, Z.; Gonzalez, F. J.; Guengerich, F. P.; Stoner, G. D.; Yang, C. S. *Cancer Res.* **1992**, *52*, 1757-1763.
13. Maser, E.; Richter, E.; Friebertshauser, J. *Eur. J. Biochem.* **1996**, *238*, 484-489.
14. Maser, E. *Trends Pharmacol. Sci.* **1997**, *18*, 270-275.
15. Maser, E. *Cancer Res.* **1998**, *58*, 2996-3003.
16. Prokopczyk, B.; Trushin, N.; Leszczynska, J.; Waggoner, S. E.; el Bayoumy, K. *Carcinogenesis* **2001**, *22*, 107-114.
17. Upadhyaya, P.; Kenney, P. M.; Hochalter, J. B.; Wang, M.; Hecht, S. S. *Carcinogenesis* **1999**, *20*, 1577-1582.
18. Wu, Z.; Upadhyaya, P.; Carmella, S. G.; Hecht, S. S.; Zimmerman, C. L. *Carcinogenesis* **2002**, *23*, 171-179.
19. London, S. J.; Daly, A. K.; Leathart, J. B.; Navidi, W. C.; Carpenter, C. C.; Idle, J. R. *Carcinogenesis* **1997**, *18*, 1203-1214.
20. Pianezza, M. L.; Sellers, E. M.; Tyndale, R. F. *Nature* **1998**, *393*, 750.
21. Hecht, S. S.; Carmella, S. G.; Murphy, S. E.; Akerkar, S.; Brunnemann, K. D.; Hoffmann, D. *N. Engl. J. Med.* **1993**, *329*, 1543-1546.
22. Hecht, S. S.; Carmella, S. G.; Ye, M.; Le, K. A.; Jensen, J. A.; Zimmerman, C. L.; Hatsukami, D. K. *Cancer Res.* **2002**, *62*, 129-134.
23. Atalla, A.; Breyer-Pfaff, U.; Maser, E. *Xenobiotica* **2000**, *30*, 755-769.
24. Oppermann, U. C. T.; Filling, C.; Hult, M.; Shafqat, N.; Wu, X.; Lindh, M.; Shafqat, J.; Nordling, E.; Kallberg, Y.; Persson, B.; Jörnvall, H. *Chem. Biol. Interact.* **2003**, *247-253*.
25. Jörnvall, H.; Persson, B.; Krook, M.; Atrian, S.; Gonzalez-Duarte, R.; Jeffery, J.; Ghosh, D. *Biochemistry* **1995**, *34*, 6003-6013.
26. Persson, B.; Kallberg, Y.; Oppermann, U. C. T.; Jörnvall, H. *Chem. Biol. Interact.* **2003**, *271-278*.
27. Sandeep, T. C.; Walker, B. R. *Trends Endocrinol. Metab.* **2001**, *12*, 446-453.
28. Seckl, J. R. *Front. Neuroendocrinol.* **1997**, *18*, 49-99.
29. Albiston, A. L.; Obeyesekere, V. R.; Smith, R. E.; Krozowski, Z. S. *Mol. Cell. Endocrinol.* **1994**, *105*, R11-R17.
30. Jamieson, P. M.; Chapman, K. E.; Edwards, C. R.; Seckl, J. R. *Endocrinology* **1995**, *136*, 4754-4761.
31. Low, S. C.; Chapman, K. E.; Edwards, C. R.; Seckl, J. R. *J. Mol. Endocrinol.* **1994**, *13*, 167-174.
32. Napolitano, A.; Voice, M. W.; Edwards, C. R.; Seckl, J. R.; Chapman, K. E. *J. Steroid Biochem. Mol. Biol.* **1998**, *64*, 251-260.
33. Voice, M. W.; Seckl, J. R.; Edwards, C. R.; Chapman, K. E. *Biochem. J.* **1996**, *317*, 621-625.
34. Maser, E.; Bannenberg, G. *Biochem. Pharmacol.* **1994**, *47*, 1805-1812.
35. Maser, E.; Oppermann, U. C. T. *Eur. J. Biochem.* **1997**, *249*, 356-369.
36. Maser, E.; Bannenberg, G. *J. Steroid Biochem. Molec. Biol.* **1994**, *48*, 257-263.

37. Maser, E.; Volker, B.; Friebertshauser, J. *Biochemistry* **2002**, *41*, 2459-2465.
38. Wermuth, B. *J. Biol. Chem.* **1981**, *256*, 1206-1213.
39. Tanaka, M.; Ohno, S.; Adachi, S.; Nakajin, S.; Shinoda, M.; Nagahama, Y. *J. Biol. Chem.* **1992**, *267*, 13451-13455.
40. Maser, E. *Biochem. Pharmacol.* **1995**, *49*, 421-440.
41. Wermuth, B.; Platt, K. L.; Seidel, A.; Oesch, F. *Biochem. Pharmacol.* **1986**, *35*, 1277-1282.
42. Maser, E.; Stinner, B.; Atalla, A. *Cancer Lett.* **2000**, *148*, 135-144.
43. Jez, J. M.; Bennett, M. J.; Schlegel, B. P.; Lewis, M.; Penning, T. M. *Biochem. J.* **1997**, *326*, 625-636.
44. Jez, J. M.; Flynn, T. G.; Penning, T. M. *Adv. Exp. Med. Biol.* **1997**, *414*, 579-589.
45. Hyndman, D.; Bauman, D. R.; Heredia, V. V.; Penning, T. M. *Chem. Biol. Interact.* **2003**, *621-631*.
46. Hara, A.; Matsuura, K.; Tamada, Y.; Sato, K.; Miyabe, Y.; Deyashiki, Y.; Ishida, N. *Biochem. J.* **1996**, *313*, 373-376.
47. Penning, T. M.; Burczynski, M. E.; Jez, J. M.; Hung, C. F.; Lin, H. K.; Ma, H.; Moore, M.; Palackal, N.; Ratnam, K. *Biochem. J.* **2000**, *351*, 67-77.
48. Deyashiki, Y.; Taniguchi, H.; Amano, T.; Nakayama, T.; Hara, A.; Sawada, H. *Biochem. J.* **1992**, *282*, 741-746.
49. Matsuura, K.; Shiraishi, H.; Hara, A.; Sato, K.; Deyashiki, Y.; Ninomiya, M.; Sakai, S. *J. Biochem. (Tokyo)* **1998**, *124*, 940-946.
50. Jin, Y.; Cooper, W. C.; Penning, T. M. *Chem. Biol. Interact.* **2003**, *383-392*.
51. Dufort, I.; Labrie, F.; Luu-The, V. *J. Clin. Endocrinol. Metab.* **2001**, *86*, 841-846.
52. Zhang, Y.; Dufort, I.; Rheault, P.; Luu-The, V. *J. Mol. Endocrinol.* **2000**, *25*, 221-228.
53. Dufort, I.; Rheault, P.; Huang, X. F.; Soucy, P.; Luu-The, V. *Endocrinology* **1999**, *140*, 568-574.
54. Breyer-Pfaff, U.; Nill, K. *Biochem. Pharmacol.* **2000**, *59*, 249-260.
55. Upadhyaya, P.; Carmella, S. G.; Guengerich, F. P.; Hecht, S. S. *Carcinogenesis* **2000**, *21*, 1233-1238.
56. Murphy, S. E.; Carmella, S. G.; Idris, A. M.; Hoffmann, D. *Cancer Epidemiol. Biomarkers & Prev.* **1994**, *3*, 423-428.
57. Ren, Q.; Murphy, S. E.; Zheng, Z.; Lazarus, P. *Drug Metab. Dispos.* **2000**, *28*, 1352-1360.
58. Breyer-Pfaff, U.; Martin, H. J.; Ernst, M.; Maser, E. *Naunyn-Schmiedebergs Arch. Pharmacol.* **2003**, *Abstract (in press)*.
59. Maser, E.; Friebertshäuser, J.; Völker, B. *Chem. Biol. Interact.* **2003**, *435-448*.
60. Higaki, Y.; Usami, N.; Shintani, S.; Ishikura, S.; El-Kabbani, O.; Hara, A. *Chem. Biol. Interact.* **2003**, *503-513*.

61. Carmines, E. L. *Food Chem. Toxicol.* **2002,** *40,* 77-91.
62. Paolini, M.; Barillari, J.; Broccoli, M.; Pozzetti, L.; Perocco, P.; Cantelli-Forti, G. *Cancer Lett.* **1999,** *145,* 35-42.
63. Monder, C.; White, P. C. *Vitam. Horm.* **1993,** *47,* 187-271.
64. Labrie, F.; Luu-The, V.; Lin, S. X.; Labrie, C.; Simard, J.; Breton, R.; Belanger, A. *Steroids* **1997,** *62,* 148-158.
65. Penning, T. M.; Pawlowski, J. E.; Schlegel, B. P.; Jez, J. M.; Lin, H. K.; Hoog, S. S.; Bennett, M. J.; Lewis, M. *Steroids* **1996,** *61,* 508-523.
66. Penning, T. M. *Endocr. Rev.* **1997,** *18,* 281-305.
67. Felsted, R. L.; Bachur, N. R. *Drug Metab. Rev.* **1980,** *11,* 1-60.

Chapter 6

Aldo-Keto Reductases and the Metabolic Activation of Polycyclic Aromatic Hydrocarbons

Trevor M. Penning[1], Nisha T. Palackal[2], Seon-Hwa Lee[1], Ian Blair[1], Deshan Yu[1], Jesse A. Berlin[3], Jeffrey M. Field[1], and Ronald G. Harvey[4]

Departments of [1]Pharmacology and [2]Biochemistry and Biophysics, and [3]Center for Clinical Epidemiology and Biostatistics, University of Pennsylvania, Philadelphia, PA 19104
[4]The Ben May Institute for Cancer Research, University of Chicago, Chicago, IL 60637

Polycyclic aromatic hydrocarbons (PAH) are suspect human lung carcinogens that are metabolically activated. Five human aldo-keto reductases [AKR1A1 (aldehyde reductase) and AKR1C1-AKR1C4] are implicated in this process since they oxidize PAH-*trans*-dihydrodiol proximate carcinogens to yield reactive and redox-active PAH-*ortho*-quinones. These *ortho*-quinones enter into futile redox-cycles to amplify reactive oxygen species (ROS) and cause DNA-lesions. AKR1C1 is highly overexpressed in non-small cell lung carcinoma patients (*1*) and in the human lung adenocarcinoma cell line A-549. AKR1C activity in these cells is sufficient to convert dimethylbenz[*a*]anthracene (DMBA)-3,4-diol to DMBA-3,4-dione. We show that AKR derived PAH *ortho*-quinones produce sufficient ROS to mutate the major lung tumor suppressor gene (*p53*) *in vitro*. These mutations eliminate the transcriptional competency of p53 in a yeast-reporter system and this change-in-function results from G to T transversions (p<0.001). These results suggest that AKRs may be involved in the etiology of human lung cancer.

Introduction

Polycyclic aromatic hydrocarbons (PAH) are tobacco carcinogens implicated in the causation of human lung cancer. Three pathways of metabolic activation have been proposed for PAH, Figure 1.

The first pathway involves the formation of radical cations by CYP peroxidase (2). Radical cations are highly electrophilic and will form depurinating adducts with DNA (3,4). The second pathway involves diol epoxide formation (e.g. *anti*-BPDE) catalyzed by the sequential actions of CYP1A1/1B1 and epoxide hydrolase (EH) (5-8). Compelling evidence exists that the diol-epoxide pathway yields ultimate carcinogens. For example, *anti*-BPDE-DNA adducts have been detected *in vitro* and *in vivo* (9-11). Diol-epoxides are the most mutagenic PAH-metabolites in the Ames test and mammalian mutagenicity assays (12) and are highly tumorigenic in the A/J mouse lung model (13,14). They also target proto-oncogenes e.g. H-*ras* to cause cell transformation (15) and adduct and mutate the tumor suppressor gene *p53* (16,17). The third pathway involves PAH *o*-quinone formation catalyzed by aldo-keto reductases (AKRs) (18); where BP-7,8-diol is converted to BP-7,8-dione. Both the diol-epoxide and *o*-quinone pathways compete for the same *trans*-dihydrodiol proximate carcinogens.

AKRs with dihydrodiol dehydrogenase activity divert *trans*-dihydrodiol proximate carcinogens from the diol-epoxide pathway to form catechols which readily autoxidize to yield reactive and redox-active *o*-quinones (19-22) Autoxidation of the catechol proceeds via two one-electron oxidations via the *o*-semiquinone anion radical producing reactive oxygen species (ROS) (23). The *o*-quinones have the ability to form covalent adducts with DNA by Michael addition (24,25). Furthermore, they can be reduced back to the catechol by either cellular reducing equivalents or by two sequential enzymatic 1 electron reductions. The PAH *o*-quinones are not substrates for quinone reductase (26). These events establish futile redox-cycles in which the formation of ROS is amplified multiple times. This oxidative stress makes a new level of genetic damage possible. This damage includes the production of oxidatively damaged bases e.g., 8-oxo-dGuo (27), DNA strand scission leading to the formation of base propenals and malondialdehyde (28), and the formation of lipid peroxides which can decompose to form bi-functional electrophiles such as 4-hydroxy-2-nonenal which can adduct DNA. In addition the creation of a prooxidant state by this pathway may be linked to tumor promotion (29). The AKR pathway provides a potential link between PAH metabolism and ROS formation, where ROS are the causative agents in radiation induced carcinogenesis. Whether this pathway of metabolic activation occurs in humans and its impact on the etiology of human lung cancer is the subject of this article.

85

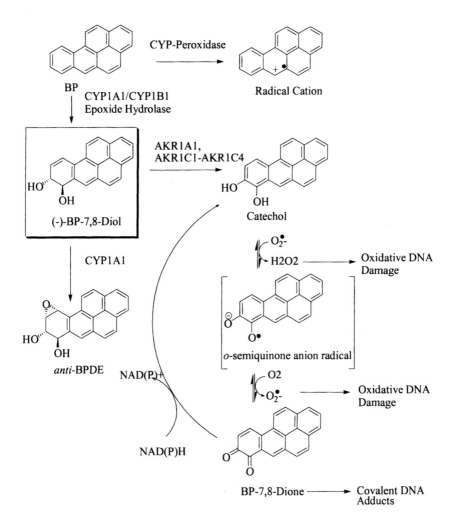

Figure 1. Metabolic routes of PAH activation.

Human Aldo-Keto Reductases and PAH *o*-Quinone Formation

Multiple human AKRs have been implicated in the formation of PAH *o*-quinones based on their ability to oxidize benzenedihydrodiol to catechol (*30-33*). These enzymes include AKR1C1-AKR1C4 which were originally purified from human liver, and share greater than 86% amino acid sequence identity. These enzymes are known by a variety of trivial names: AKR1C1 is also 20α-hydroxysteroid dehydrogenase (HSD) and dihydrodiol dehydrogenase 1; AKR1C2 is also type 3 3α-HSD, bile-acid binding protein and dihydrodiol dehydrogenase 2; AKR1C3 is also type 2 3α-HSD, type 5 17β-HSD and dihydrodiol dehydrogenase X; and AKR1C4 is type 1 3α-HSD, dihydrodiol dehydrogenase 4 and chlordecone reductase. In addition, AKR1A1 (human aldehyde reductase) may be involved in PAH *o*-quinone formation (*34*). This latter enzyme catalyzes the reduction of glyceraldehyde to glycerol and mevaldate to mevalonic acid. This enzyme is thus involved in triglyceride and cholesterol metabolism. The possibility that this general metabolic enzyme may be involved in PAH activation is provocative.

Involvement of Aldehyde Reductase (AKR1A1) in PAH Activation

To determine whether AKR1A1 may catalyze the AKR pathway of PAH activation, the cDNA for AKR1A1 was obtained by RT-PCR using poly(A)$^+$-RNA from human hepatoma cells (HepG2) as template (*21*).

The cDNA was subcloned into a prokaryotic expression vector (pET16b) and the recombinant protein was purified from the *E. coli* host in milligram amounts. Recombinant AKR1A1 catalyzed the NADP$^+$-dependent oxidation of (±)-*trans*-7,8-dihydroxy-7,8-dihydrobenzo[*a*]pyrene (BP-7,8-diol) with a utilization ratio (V_{max}/K_m) superior to any other human isoform (also, see Table 1). RP-HPLC analysis indicated that only 50% of the racemic BP-7,8-diol was oxidized. Isolation of the unreacted isomer followed by CD-spectrometry showed that the (+)-*S,S*-isomer remained, while the isomer formed metabolically *in vivo*, the (-)-*R,R*-diastereomer was preferentially oxidized, Figure 2. Thus, AKR1A1 displays the correct stereochemistry to be involved in the *in vivo* metabolism of BP-7,8-diol.

The product of the AKR1A1 reaction was identified as the highly reactive BP-7,8-dione which was trapped as its thio-ether conjugate with 2-mercaptoethanol. The thio-ether conjugate was characterized by LC/MS yielding a MH$^+$ (*m/z* = 359) and a fragment ion corresponding to the loss of 1 carbonyl group [MH$^+$-CO; *m/z* = 331]. The spectrum of the conjugate was identical to that observed for the authentic synthetic standard. The product peak could be easily distinguished from the substrate BP-7,8-diol which gave a MH$^+$ minus water [MH$^+$-H$_2$O; *m/z* = 269].

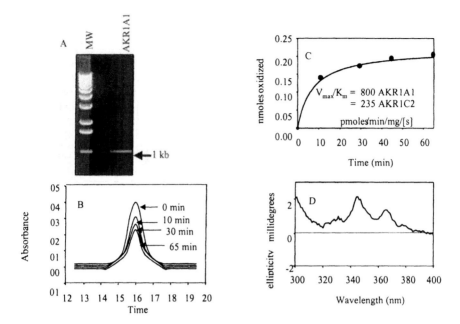

Figure 2. Role of aldehyde reductase in the metabolism of (-)-BP-7,8-diol. RT-PCR cloning from HepG2 cell total RNA (A); HPLC time course for the NADP+ dependent oxidation of BP-7,8-diol mediated by recombinant AKR1A1 (B); initial velocity for BP-7,8-diol oxidation with comparison of Vmax/Km for AKR1C2 (C); and CD-spectra showing that the (+)-BP-7,8-diol isomer remains at the end of the AKR1A1 catalyzed reaction (D).

Because tissues are exposed to the parent PAH rather than the *trans*-dihydrodiol substrate, we determined whether AKR1A1 was expressed in PAH target tissues and whether there was co-expression with CYP1A1 and EH. Multiple human tissue expression array analysis was performed. All three enzymes were expressed in liver, salivary gland, trachea, lung and esophagus; all sites of PAH exposure, suggesting that the enzymatic machinery exists to activate PAH to PAH *o*-quinones in these tissues via AKR1A1.

Involvement of AKR1C1-AKR1C4 in PAH Activation

The remaining AKR isoforms (AKR1C1-AKR1C4) implicated in PAH metabolism/activation were cloned and expressed in a manner similar to that described for AKR1A1 (*35*). SDS-PAGE and immunoblot analysis verified that each enzyme was obtained in homogeneous form. The utilization ratios for the oxidation of a structural series of *trans*-dihydrodiols: BP-7,8-diol (bay-region); dimethylbenz[*a*]anthracene-3,4-diol (methylated bay region) and benz[*g*]chrysene-11,12-diol (*fjord*-region) were determined using recombinant AKR1C1-AKR1C4. Structure-activity relationships show that the carcinogenicity of diol-epoxides increases when the bay region is methylated or when it becomes more hindered due to the presence of a *fjord* region. To determine whether AKRs could oxidize the *trans*-dihydrodiols of these potent proximate carcinogens reactions were performed with the recombinant enzymes. Aldehyde Reductase (AKR1A1) had the highest utilization ratio for BP-7,8-diol while AKR1C4 had the highest utilization ratio for methylated and *fjord*-region *trans*-dihydrodiols, Table 1 (*22,35,36*). In examining the stereochemical preference for each substrate, AKR1A1 was stereospecific while AKR1C1-AKR1C4 consumed both isomers of the racemic substrates.

Table 1. Oxidation of PAH *trans*-dihydrodiols by human AKR Isoforms

PAH *trans*-dihydrodiol	AKR1A1	AKR1C1	AKR1C2	AKR1C3	AKR1C4
	V_{max}/K_m (pmoles^{-1}min^{-1}/[S])				
Benzo[*a*]pyrene-7,8-diol	800	180	235	22	103
Dimethylbenz[*a*]anthracene-3,4-diol	2625	200	1266	531	5000
Benzo[*g*]chrysene-11,12-diol	304	122	133	620	4400

AKRs and the Etiology of Human Lung Cancer

If AKRs oxidize PAH *trans*-dihydrodiols to *o*-quinones what is their relationship to human lung cancer? Using the technique of differential display it was found that AKR1C1 was overxpressed by up to 50-fold in 83% of patients with non-small cell lung carcinoma (*1*). If PAH *o*-quinones are produced in lung carcinoma cells can they also account for the most common mutations (G to T transversions) in the proto-oncogene *K-ras* (*37*) and in the tumor suppressor gene *p53* (*38*) which are among the most commonly mutated genes in lung cancer? G to T transversions occur at discrete sites known as "hot-spots" in these genes and cause a change in function resulting in altered cell growth properties.

Functional AKR1C Expression in Human Lung Adenocarcinoma

To validate the results from the differential display experiments, microarray analysis was performed on 96 human tissues and cells using a cDNA probe for AKR1C1 (*22*). Highest expression levels were observed in the human lung adenocarcinoma cell line A549. Northern analysis using RNA from the A549 lung carcinoma cell line confirmed that AKR1C1 transcript was overexpressed, Figure 3.

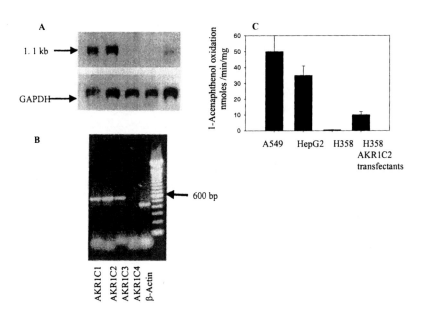

Figure 3. Overexpression of AKR1C isoforms in human lung A549 cells. (A) Northern analysis showing 1.1 kb transcript in HepG2 cells (lane 1); A549 cells (lane 2); H441 cells (lane 3); H358 cells (lane 4); and H358-AKR1C2 transfectants (lane 5); (B) RT-PCR detection of transcripts in A549 cells; and (C) enzyme activity measurements in A549 cells using 1-acenaphthenol as substrate.

Subsequently, isoform specific RT-PCR showed that AKR1C1-AKR1C3 transcripts were all overexpressed in A549 cells whereas AKR1C4 was not because it is liver specific. Using 1-acenaphthenol as an artificial substrate, higher enzyme activity was observed in the A549 cells than that observed in human hepatoma cells, and 50-fold greater activity was observed than that seen in human bronchoalveolar cells of epithelial origin (H-358). The activity was also 5-fold greater activity than that observed in H-358 cells stably transfected with AKR1C2.

PAH *o*-Quinone Formation in Lung Adenocarcinoma Cells

To determine whether the activity in human A549 cells is sufficient to make PAH *o*-quinones, A549 cell lysates were challenged with the potent proximate carcinogen DMBA-3,4-diol. Because DMBA-3,4-dione is one of the most reactive PAH *o*-quinones that can be produced by the AKR pathway, attempts were again made to trap the cell-derived product as its thioether conjugate. It was found that DMBA-3,4-dione was trapped as its *mono* and *bis*-thioether conjugates with 2-mercaptoethanol. LC/MS analysis of the *mono*-conjugate gave MH$^+$ (m/z = 363) and a fragment ion of [MH-H$_2$O$^+$; m/z = 345]. By contrast, LC/MS of the *bis*-conjugate gave a MH$^+$ (m/z= 439) and fragment ions of [MH-H$_2$O$^+$, m/z = 421; and MH- SCH$_2$CH$_2$OH$^+$, m/z = 362] were observed. In each instance the spectra were identical to the authentic synthetic standard. [^{1}H]-NMR of the *mono*-conjugate shows the absence of the vinylic proton at C2. This supports a mechanism in which there is sequential 1,6- followed by 1,4-Michael addition of the thiol to yield the final *bis*-conjugate (*22*). Between each addition there would be autoxidation of the intermediate catechol with the concomitant production of ROS.

DNA-Lesions Caused by PAH *o*-Quinone Exposure

PAH *o*-quinones produced by the AKR pathway have the potential to produce a variety of DNA-lesions that could cause G to T transversions in target genes e.g., *p53* and *K-ras*.

First, they may form covalent adducts by 1,4-Michael addition to form stable-bulky adducts with either the N^2- exocyclic amino group or the N^6-exocyclic amino groups of deoxyguanosine and deoxyadenosine, respectively. Translesional synthesis by error-prone by-pass DNA polymerases may then give rise to G to T transversions, Figure 4.

Evidence for stable adducts exists. *In vitro* the reaction of [^{3}H]-BP-7,8-dione with calf thymus DNA followed by subsequent digestion of the DNA to its constituent deoxyribonucleosides gave a single adduct which co-migrated with that formed by the reaction of [^{3}H]-BP-7,8-dione with oligo-dG followed by enzymatic digestion (*24*), Figure 5.

Second, PAH *o*-quinones could undergo nucleophilic attack by N7 of a purine base with resultant cleavage of the glycosidic bond to give BP-7,8-dione-

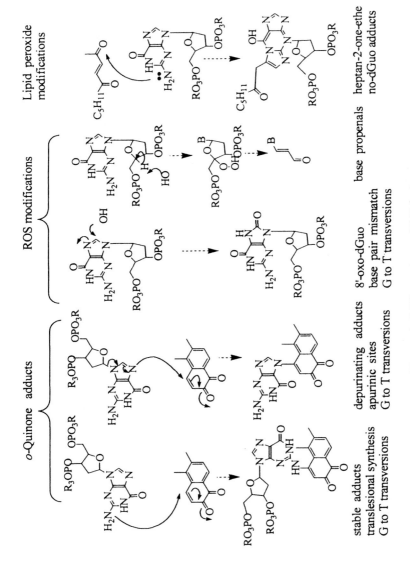

Figure 4. PAH o-quinone-DNA lesions.

N7-guanine, or BP-7,8-dione-N7-adenine depurinating adducts. The resultant abasic site has a preference for the introduction of A opposite that site giving rise to a G to T transversion on the daughter strand. Depurinating adducts for naphthalene-1,2-dione, phenanthrene-1,2-dione and benzo[*a*]pyrene-7,8-dione have been synthetically prepared and detected upon treatment of calf thymus DNA with PAH *o*-quinones. In each instance the depurinating adducts were fully characterized by LC/MS (*25*).

Third, because PAH *o*-quinones are redox-active they can produce ROS. ROS attack of deoxyguanosine can give rise to 8-oxo-deoxyguanosine which can be detected by EC-HPLC following digestion of the exposed DNA. High levels of 8-oxo-dGuo have been detected in calf thymus DNA following treatment with naphthalene-1,2-dione under redox cycling conditions (e.g. NADPH and S9), (> 200 8-oxo-dGu/10^5 dG (*26,27*). If unrepaired, 8-oxo-dGuo will mispair with adenine, providing an additional straightforward route to a G to T transversion.

Fourth, ROS production can lead to hydroxyl radical attack of deoxyribose leading to DNA strand scission and the formation of base propenals which hydrolyze to yield malondialdehyde, which is highly mutagenic (*28*). DNA-strand scission mediated by PAH *o*-quinones has been observed under redox-cycling conditions using poly-dC.dG, ϕX174 DNA and in rat hepatocytes (*29*). Apart from ROS modifications of DNA, ROS can also attack polyunsaturated fatty acids to yield lipid hydroperoxides which can decompose to yield bi-

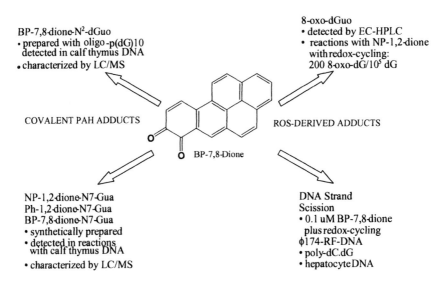

Figure 5. DNA-lesions detected with PAH o-quinones.

functional electrophiles, e.g. 4-hydroxy-2-nonenal and 4-oxo-2-nonenal which can give rise to etheno-adducts of unknown mutagenic potential. Thus, PAH *o*-quinones can cause a variety of DNA-lesions that could result in G to T transversions. The question exists as to whether PAH *o*-quinones can cause mutations in critical target genes.

ROS Generated by PAH *o*-Quinones Cause Change-in-Function Mutations in *p53*.

The *p53* tumor suppressor gene is mutated in 60% of all human lung cancers. In smokers "hot-spots" undergo point mutations and 76% are G to T transversions (*38*). Susceptible codons that cause a change-in-function in the DNA-binding domain of *p53* include 157, 158, 245, 248, 249 and 275. We have established a yeast-reporter system that measures changes in the transcriptional competency of p53 following treatment with a mutagen (*39*). In this assay wt p53 cDNA is treated with a mutagen, and the cDNA is inserted back into a "gap-repair" vector by homologous recombination upon transformation into yeast strains. The host yeast strain contains an adenine reporter gene stably integrated into its chromosome under the control of the p21 promoter. In the presence of limiting adenine wt p53 will turn yeast colonies white, while mutated p53 will turn colonies red (positive selection), Figure 6.

As a positive control WT p53 was treated with the test mutagen, *N*-methyl-*N*-nitroso-*N*-nitroguanidine. Isolation of the p53 plasmid from the red colonies followed by re-transformation validated that a change-in-function mutation was present. Dideoxysequencing revealed that 15/17 mutations were C to T transitions, this high preponderance of this single point mutation is consistent with previous findings with this alkylating mutagen (*40*).

In this assay BP-7,8-dione was found to be ineffective by itself or in combination with NADPH. However, BP-7,8-dione was found to be a potent p53 mutagen at sub-micromolar concentrations in the presence of NADPH and $CuCl_2$ to amplify the production of ROS. At these concentrations ds-cDNA for p53 was not fragmented. Thus, at 0.25 μM BP-7,8-dione the mutation frequency observed was 5.1% (113/2200 colonies). A similar mutagenic frequency was seen with (±) *anti*-BPDE at concentrations that were 80x greater. The mutagenicity of BP-7,8-dione was attenuated by ROS scavengers but completely blocked by catalase and superoxide dismutase indicting that both OH• radical and superoxide anion were the responsible mutagens.

The mutational pattern caused by BP-7,8-dione in p53 resembled that found in human lung cancer. The p53 plasmids were isolated from the mutated colonies and the DNA-binding domain was subjected to dideoxysequencing. It was found that the ROS produced by BP-7,8-dione caused single-point mutations rather than multiple mutations. The frequency of G to T transversions

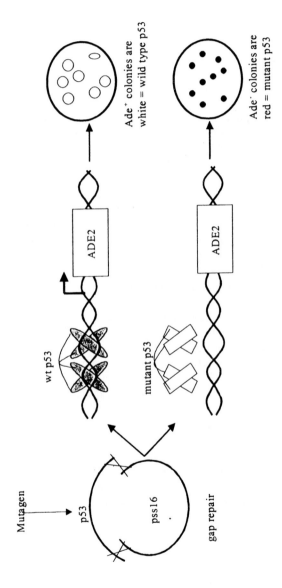

Figure 6. p53 Mutation is scored in a yeast reporter gene assay.

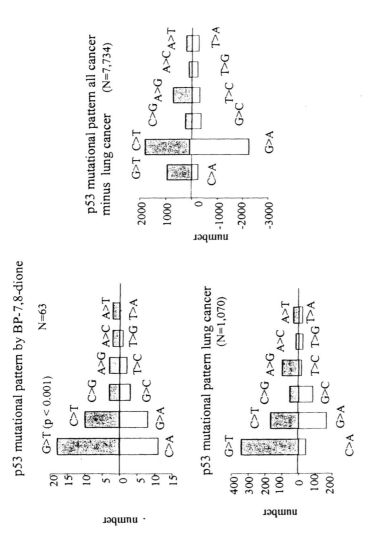

Figure 7. Mutation of p53 by PAH o-Quinones under redox cycling conditions.

versus the other 12 possible mutations (1 base mutated to each of the remaining three) were compared. Since the assay does not distinguish whether the initial lesion occurred on the coding or non-coding strand the number of possible mutations is reduced to six. Power calculations were performed to determine the number of colonies that would have to be sequenced to confirm that the incidence of G to T transversions was not random (n = 83). Upon analyzing 63 mutants it was found that G to T transversions occurred 46% of the time yielding a p <0.001. This high level of G to T transversions that are dependent upon ROS would be consistent with the formation of 8-oxo-dGuo, Figure 7.

This mutational pattern mirrors that seen in lung cancer from the p53 mutational data base and is distinctly different from that seen in other cancers where C to T transitions dominate (*38*). Although there is an insufficient n value to develop a mutational spectrum (incidence of single point mutation by codon number), 25% of the mutations detected were at hotspots in p53 which are known to be mutated in lung cancer (p < 0.007). When our data are combined with that seen with *anti*-BPDE two non-exclusive paradigms may contribute to the etiology of lung cancer, Figure 8.

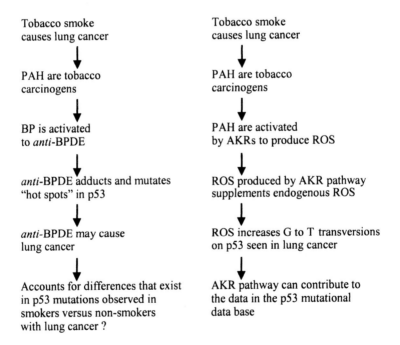

Figure 8. Contribution of AKR pathway to mutations in lung cancer.

In the first paradigm, we know that tobacco smoke causes lung cancer and PAH are tobacco carcinogens. Benzo[*a*]pyrene is a representative PAH and is activated to (±)-*anti*-BPDE. *Anti*-BPDE forms adducts and mutates *p53* at hotspots. Therefore, (±)-*anti*-BPDE may contribute to the causation of lung cancer. This would account for the differences that exist in p53 mutations observed in smokers versus non-smokers. In the second paradigm, PAH which are tobacco carcinogens are activated by AKRs in human lung cells to produce ROS. The ROS produced by AKRs supplements the endogenous ROS formed. ROS increases the incidence of G to T transversions in p53 seen in lung cancer. The AKR pathway may contribute to the data in the p53 mutational database as it relates to lung cancer.

Acknowledgements

This work was supported by NIH grants, R01-CA3904 and P01-92537, awarded to T.M.P.

References

1. Hsu, N-Y.; Ho, H-C.; Chow, K-C.; Lin, T-Y.; Shih, C-S.; Wang, L-S.; Tsai, C-M. *Cancer Res.* **2001**, *61,* 2727-2731.
2. Cavalieri, E. L.; Rogan, E. G. *Xenobiotica* **1995**, *25,* 677-688.
3. Devanesan, P. D.; RamaKrishna, N. V. S.; Todorovic, R.; Rogan, E. G.; Cavalieri, E. L.; Jeong, H.; Jankowiak, R.; Small, G. J. *Chem. Res. Toxicol.* **1992**, *5,* 302-309.
4. Chen, L.; Devanesan, P. D.; Higginbotham, S.; Ariese, F.; Jankowiak, R.; Small, G. J.; Rogan, E. G.; Cavalieri, E. *Chem. Res. Toxicol.* **1996**, *9,* 897-903.
5. Gelboin, H. V. *Physiol. Rev.,* **1980**, *60,* 1107-1166.
6. Conney, A. H. *Cancer Res.* **1982**, *42,* 4875-4917.
7. Shimada, T.; Martin, M. V.; Pruess-Schwartz, D.; Marnett, L. J.; Guengerich, F. P., *Cancer Res.* **1989**, *49,* 6304-6312.
8. Shimada, T.; Hayes, C. L.; Yamazaki, H.; Amin, S.; Hecht, S. S.; Guengerich, F. P.; Sutter, T. R. *Cancer Res.* **1996**, *56,* 2979-2984.
9. Jennette, K. W.; Jeffery, A. M.; Blobstein, S. H.; Beland, F. A.; Harvey, R. G.; Weinstein, I. B. *Biochemistry* **1977**, *16,* 932-938.
10. Jeffrey, A. M.; Jennette, K. W.; Blobstein, S. H.; Weinstein, I. B.; Beland, F. A.; Harvey, R. G.; Kasai, H.; Miura, I.; Nakanishi, K. *J. Amer. Chem. Soc.* **1976**, *98,* 5714-5715.
11. Koreeda, M.; Moore, P. D.; Wislocki, P. G.; Levin, W.; Conney, A. H.; Yagi, H.; and Jerina, D. M. *Science* **1978**, *199,* 778-781.

98

12. Malaveille, C.; Kuroki, T.; Sims, P.; Grover, P. L.; Bartsch, H. *Mutat. Res.* **1977**, *44*, 313-326.
13. Buening, M. K.; Wilsocki, P. G.; Levin, W.; Yagi, H.; Thakker, D. R.; Akagi, H.; Koreeda, M.; Jerina, D. M.; Conney, A. H. *Proc. Natl. Acad. Sci. USA* **1978**, *75*, 5358-5361.
14. Chang, R. L; Wood, A. W.; Conney, A. H.; Yagi, H.; Sayer, J. M.; Thakker, D. R.; Jerina, D. M.; Levin, W. *Proc. Natl. Acad. Sci. USA* **1987**, *84*, 8633-8636.
15. Marshall, C. J.; Vousden, K. H.; Phillips, D. H. *Nature* **1984**, *310* 585-589.
16. Denissenko, M. F.; Pao, A.; Tang, M-S.; Pfieifer, G. P. *Science* **1996**, *274*, 430-432.
17. Hussain, S. P.; Amstad, P.; Raja, K.; Sawyer, M.; Hofseth, L.; Shields, P.; Hewer, A.; Phillips, D. H.; Ryberg, D.; Huagen, A.; Harris, C. C. *Cancer Res.* **2001**, *61*, 6350-6355.
18. Penning, T. M.; Burczynski, M. E.; Hung, C-F.; McCoull, K. D.; Palackal, N. T.; Tsuruda, L. S. *Chem. Res. Toxicol.* **1999**, *12*, 1-18.
19. Smithgall, T. E.; Harvey, R. G.; Penning, T. M. *J. Biol. Chem.* **1986**, *26*, 6184-6191.
20. Smithgall, T. E.; Harvey, R. G.; Penning, T. M. *J. Biol. Chem.*, **1988**, *263*, 1814-1820.
21. Palackal, N. T.; Burczynski, M. E.; Harvey, R. G.; Penning, T. M. *Biochemistry* **2001**, *40*, 10901-10910.
22. Palackal, N. T.; Lee, S-H.; Harvey, R. G.; Blair, I. A.; Penning, T. M. *J. Biol. Chem.* **2002**, *277*, 24799-24808.
23. Penning, T. M.; Ohnishi, S. T.; Ohnishi, T.; Harvey, R. G. *Chem. Res. Toxicol.* **1996**, *9*, 84-92.
24. Shou, M.; Harvey, R. G.; Penning, T. M. *Carcinogenesis* **1993**, *14*, 475-482.
25. McCoull, K. D.; Rindgen, D.; Blair, I. A.; Penning, T. M. *Chem. Res. Toxicol.* **1999**, *12*, 237-246
26. Flowers-Geary, L.; Harvey, R. G.; Penning, T. M. *Biochem. (Life. Sci.Adv.)* **1992**, *1,1* 49-58.
27. Glaze, E. R.; Flowers, L.; Penning, T. M. *Proc. Amer. Assoc. Cancer Res.* **2001**, *42*, 2542.
28. Flowers, L.; Ohnisni, S. T.; Penning, T. M. *Biochemistry* **1997**, *36*, 8640-8648.
29. Cerutti, P. A. *Science* **1985**, *227*, 375-381.
30. Deyashiki, Y.; Taniguchi, H.; Amano, T.; Nakayama, T.; Hara, A.; Sawada, H. *Biochem. J.* **1992**, *282*, 741-746.
31. Stolz, A.; Hammond, L.; Lou, H.; Takikawa, H.; Ronk, M.; Shively, J. E. *J. Biol. Chem.* **1993**, *268*, 10448-10457.
32. Deyashiki, Y, Tamada, Y., Miyabe, Y., Nakanishi, M., Matsuura, K., and Hara, A. J. *Biochem.* **1995**, *118*, 285-290.
33. Hara, A.; Matsurra, K.; Tamada, Y.; Sato, K.; Miyabe, Y.; Deyashiki, Y.; Ishida, T. *Biochem. J.* **1996**, *313*, 373-376.

34. Hara, A.; Taniguchi, H.; Nakayama, T.; Sawada, H. *J. Biochem.* **1990,** *108,* 250-254.
35. Burczynski, M E.; Harvey, R.G.; Penning, T.M. *Biochemistry* **1998,** *37,* 6781-6790.
36. Palackal, N. T.; Burczynski, M. E.; Harvey, R. G.; Penning, T. M. *Chemico. Biol. Inter.* **2001,** *130-132,* 815-824.
37. Rodenhius, S. *Cancer Biology* **1992,** *3,* 241-247.
38. Soussi, T. *The p53 Database.* http://p53.curie.fr/p53%20site%20 version%202.0/database/p53_database.html, 2001.
39. Yu, D.; Berlin, J. A.; Penning, T. M.; Field, J. *Chem. Res. Toxicol.* **2002,** *15,* 832-842.
40. Koch, W. H.; Henrikson, E. H.; Kopechella, E.; Cebula, T. A.; *Carcinogenesis* **1994,** *15,* 79-88.

Chapter 7

Molecular Cloning and Characterization of Dihydrodiol Dehydrogenase from the Mouse

Yoshihiro Deyashiki[*], Takahiro Takatsuji, and Akira Hara

Laboratory of Biochemistry, Gifu Pharmaceutical University,
5–6–1, Mitahora-higashi, Gifu 502–8585, Japan

A full-length cDNA encoding a novel protein has been cloned by screening a mouse liver cDNA library with human dihydrodiol dehydrogenase isoform (AKR1C1) cDNA. The cDNA contained an open reading frame of mouse dihydrodiol dehydrogenase (AKR1C22) consisting of 323 amino acids. AKR1C22 exhibited more than 85% amino acid sequence with the aldo-keto reductase superfamily proteins from mouse (AKR1C12, AKR1C13) and rat (AKR1C16, AKR1C17). Recombinant AKR1C22 oxidized benzene dihydrodiol and reduced various carbonyl compounds including isatin and diacetyl, and preferred NAD(H) to NADP(H) as the coenzymes. The dehydrogenase and reductase activities of AKR1C22 were inhibited by hexestrol and medroxyprogesterone. mRNA for AKR1C22 was predominantly expressed in liver and small intestine. The results indicate that AKR1C22 is a new member of the aldo-keto reductase superfamily and may be involved in the detoxification of xenobiotics and in the metabolism of some endogenous carbonyl compounds.

Introduction

Dihydrodiol dehydrogenase (DD; EC 1.3.1.20) catalyzes $NADP^+$-linked oxidation of *trans*-dihydrodiols of aromatic hydrocarbons to corresponding catechols. The enzyme has dual toxicological roles in the metabolism of polycyclic aromatic hydrocarbons. The enzyme suppresses the formation of their carcinogenic dihydrodiol epoxides (*1,2*), while it is also involved in the production of cyto- and genotoxic *o*-quinones, which can react with cellular nucleophiles (GSH, RNA and DNA) (*3-5*), through auto-oxidation of the catechol intermediates.

We previously purified four forms (DD1-DD4) of monomeric DD from human liver, and demonstrated that three forms (DD1, DD2 and DD4) showed distinct specificity for 3α- and/or 20α-hydroxysteroids and prostaglandins (*6-9*). Based on the nomenclature for the aldo-keto reductase (AKR) superfamily (*10, 11*) and the amino acid sequences deduced from their cDNAs (*9,12,13*), DD1, DD2 and DD4 have been classified to AKR1C1, AKR1C2 and AKR1C4, respectively. We also showed that homogeneous AKR1C3 displayed dihydrodiol dehydrogenase activity (*14*). The four DDs (AKR1C1-4) belong to the AKR1C subfamily, the structural features of which are well characterized by the analysis of the crystal structures of rat AKR1C9 (*15*) and human AKR1C2 (*16*). Recently, the activation of polycyclic aromatic hydrocarbons by the enzymes (AKR1C1-4) has been shown *in vitro* by Penning et al. (*17,18*). To date, the correspondence of human AKR1C enzymes to DDs (*19-21*) purified from mouse is unclear at the nucleotide sequence level.

The mouse is a useful model for investigating the metabolism of dihydrodiols and carbonyl compounds. We describe the cloning, expression and preliminary characterization of a murine dihydrodiol dehydrogenase (AKR1C22) that prefers NAD(H) to NADP(H).

Experimental Procedures

Isolation of cDNA Clones and DNA Sequencing

A mouse (adult male, Balb/c strain) liver 5'-strech cDNA library in λgt 11 vector (catalogue no. ML1035b) was purchased from Clontech Laboratories (USA). The library (2.7×10^5 plaques) was screened by plaque hybridization by using biotinylated cDNA corresponding to the open reading frame of human liver DD1 (AKR1C1) cDNA (*9,12*) as a probe. The biotinylated cDNA probe was prepared by polymerase chain reaction (PCR) with primers used for recombinant human liver DD1 expression (*9*) using biotinylated dATP (Gibco

BRL) as an additional substrate (22). Hybridization was carried out at 65 °C in 5x SSPE (1x SSPE: 0.18 M NaCl, 10 mM sodium phosphate, pH 7.7, and 1 mM EDTA), 0.5% sodium dodecyl sulfate (SDS), 5x Denhardt's solution and 0.02 mg/ml denatured and sheared salmon sperm DNA, and filters were washed twice with 2x SSPE at 65 °C for 15 minutes and once with 2x SSPE containing 0.1% SDS at 65 °C for 30 minutes. After washing, filters were treated at 60 °C for an hour with 100 mM Tris-HCl, pH 7.5, containing 1.0 M NaCl, 2 mM $MgCl_2$, 0.05% Triton X-100 and 3% bovine serum albumin, and bound biotinylated probe was detected with nitro blue tetrazolium, 5-bromo-4-chloroindolyl phosphate and streptavidin-conjugated alkaline phosphatase. Positive phage were isolated and purified. The inserts were digested with *Eco R*I and subcloned into the pBluescript SK(-) vector for DNA squencing.

Expression and Purification of Recombinant AKR1C22 Proteins

Three oligonucleotides: mDDN1, 5'-CCGGATCCGATGAGCTCCAAA CAGCACTG-3', mDDN2, 5'-AGCATATGAGCTCCAAACAGCAC-3' and mDDCX, 5'-CCCTCGAGTTAATATTCCTCCACAAATGG-3' were synthesized. mDDN1 and mDDN2 corresponded to nucleotides 1 - 20 and 1 - 18, respectively, of the AKR1C22 cDNA contained a *Bam* HI and a Nde I site, respectively. The mDDCX primer was complementary to nucleotides 952 - 972 of the cDNA and contained a *Xho* I site. The DNA fragments to create either the histidine-tag (His-tag) AKR1C22 fusion protain or AKR1C22 were amplified by PCR with *Pfu* DNA polymerase using mDDN1 and mDDCX, and mDDN2 and mDDCX, respectively, and the cloned cDNA as a template. The amplified cDNA fragments were inserted into the corresponding cloning sites of the pRSET plasmids. The resultant plasmid DNAs were sequenced to verify their fidelity and transfected into the host strain *E. coli* BL21(DE3). Expression of recombinant protein and preparation of the cell crude extract were prepared as previously described (9).

Recombinant His-tag fusion AKR1C22 was purified to homogeneity from the crude extract of transfected cells by TALON (QIAGEN), an immobilized metal affinity resin, column chromatography. Recombinant AKR1C22 was purified to homogeneity by four steps consisting of 35-85% ammonium sulfate precipitation, and Sephadex G-100, Q-Sepharose and Matrex Red A column chromatographies. Homogeneity was confirmed by sodium dodecyl sulfate-polyacryl amide gel electrophoresis (SDS-PAGE) and staining with Quick-CBB (Wako Pure Chemicals, Osaka, Japan).

The masses of recombinant enzyme and its denatured enzyme were estimated by Sephadex G-100 filtration and SDS-PAGE, respectively. Protein concentration was determined using bovine serum albumin as the standard by the method of Bradford (23).

Enzyme Assay

Dehydrogenase and reductase activities of the recombinant enzymes were assayed by recording the production and oxidation, respectively, of NAD(P)H as described (6). The standard reaction mixture for the dehydrogenase activity consisted of 0.1 M potassium phosphate, pH 7.4, 0.25 mM NAD(P)$^+$, 1 mM S-tetralol, and enzyme in a total volume of 2.0 ml. The reductase activity was determined with 2.0 ml of 0.1 M potassium phosphate, pH 7.4, 0.1 mM NAD(P)H, carbonyl substrates, and enzyme. One unit of the enzyme activity was defined as the amount of enzyme catalyzing the formation or oxidation of 1 μmol NAD(P)H/min at 25 °C. For determining the effect of pH on the activity, we used 100 mM potassium phosphate buffer (pH 5.0-10.5). Benzene dihydrodiol was synthesized as described by Platt and Oesch (24).

RT-PCR of AKR1C22 mRNA

Total RNA from mouse tissues (brain, heart, kidney, liver, lung, small intestine, spleen, stomach and blood) was isolated using ISOGEN (Nippon Gene, Tokyo, Japan) and reverse transcribed using Moloney murine leukemia virus revers transcriptase and an oligo (dT)$_{12\text{-}18}$ primer. Three nucleotides: mDD-sp, 5'-GAGCAAGGGAAATCTCTGTTG-3', AKR1C6-sp, 5'-CTAAGCAGCAGACAGTGCG-3' and AKRr, 5'-CTTGTTCTTTTTGGCCACATCA-3' were synthesized. mDD-sp and AKR1C6-sp corresponded to nucleotides 403 - 424 of AKR1C22 cDNA and 15 - 33 of mouse 17β-hydroxysteroid dehydrogenase (AKR1C6) cDNA, respectively. The AKRr primer was complementary to nucleotides 732 - 754 of AKR1C22 and AKR1C6. Polymerase chain reaction (PCR) was carried out with mDD-sp and AKRr, and AKR1C6-sp and AKRr, to detect mRNAs for AKR1C22 and AKR1C6, respectively.

Results and Discussion

Cloning and cDNA Sequence of AKR1C22

A mouse liver cDNA library was screened by plaque hybridization with human DD isoform (AKR1C1) cDNA as a probe. Seven positive clones were isolated from 2.7 x 10^5 plaques. Two clones contained a full-length cDNA encoding a novel protein, which is named mouse DD (mDD, AKR1C22), consisting of 323-amino acid residues (Figure 1). Three other clones contained a cDNA identical with AKR1C6. The other two clones were predicted to encode the COOH-terminal part of an unkown protein that is similar to AKR1C6. The full-length cDNA of AKR1C22 was 1191 bp long and contained a 6 bp 5'-flanking region, a 972 bp open reading frame encoding a 323 amino acid polypeptide, and a 213 bp 3'-untransrated region with a poly(A) tail (Figure 1). The molecular weight of the deduced protein was estimated as 37,042.

Amino acid sequence analysis revealed that AKR1C22 showed significant identity to known proteins belonging to the aldo-keto reductase (AKR) superfamily. The amino acid sequence of AKR1C22 was 99%, 94%, 91%, 85%, 65%, 65% and 64% identical with those of mouse aldo-keto reductase AKR1C12 (26), mouse interleukin-3-regulated AKR (AKR1C13) (26, 27), rat AKRB (AKR1C16) (28), rat AKRD (AKR1C17) (28), human liver DD4 variant (AKR1C4) (29), mouse 17β-HSD (AKR1C6) (25) and rat 3α-HSD (AKR1C9) (30), respectively. AKR1C22 differed by two residues (Arg47 and Val81) from the sequence of AKR1C12 as shown in Figure 2. Moreover, its amino acid sequence was identical to that of BC021937 (accession number of GenBank/EMBL) and its nucleotide sequence (1191 bases) was identical to the part of BC021937 (1224 bases). In addition, AKR1C22 also shared high identity (84%) with hamster morphine 6-dehydrogenase (M-6-D) (31) when 303 of their sequenced amino acid residues were aligned (Figure 2). In the AKR1C22 sequence, 4 residues (Asp50, Tyr55, Lys84, and His117) involved in catalysis of aldo-keto reductases were conserved. The results indicate that AKR1C22 belongs to the AKR superfamily and that it could be a new member of the AKR1C subfamily.

Of the 9 residues involved in the binding of ursodeoxycholate to AKR1C2 (16), AKR1C22 contained Tyr24, Val54, Tyr55, Trp86, His117 and Trp227, but Val128, Ile129, and Leu308 in the AKR1C2 sequence were substituted with Asp, Phe, and Ala, respectively. Only 3 residues (Thr24, Leu54, and Phe128) involved in the substrate binding of rat 3α-HSD (15) were substituted with Tyr24, Val54, and Asp128 in the AKR1C22 sequence. The conservation of active-site residues (Asp50, Tyr55, Lys84, and His117) in AKR1C22, suggest that the catalytic mechanism of AKR1C22 is the same as other enzymes in the AKR1C subfamily.

Expression of His-tag fusion AKR1C22 and AKR1C22

Recombinant His-tag fusion AKR1C22 and AKR1C22 were expressed in *E. coli* (BL21(DE3)) and purified to homogeneity based on SDS-PAGE. From 500 ml of bacterial culture, approxymately 7 and 4.7 mg of recombinant His-tag fusion AKR1C22 and AKR1C22, respectively, were obtained. The specific activities of His-tag fusion AKR1C22 and AKR1C22 were 0.6 and 0.95 U/mg protein, respectively, in the assay of dehydrogenase activity with 1 mM *S*-tetralol and 0.25 mM NAD^+ in 0.1 M glycine-NaOH at pH 10. The molecular weight (37,000) of denatured AKR1C22 determined by SDS-PAGE was the same as the

GAGACAATGAGCTCCAAACAGCACTGTGTCAAACTAAATGATGCCACTTAATTCCTCGCCCTGGCTTTGGCACCTATAAACCCAAGGAG   90
　　　　　　M　S　S　K　Q　H　C　V　K　L　N　D　G　H　L　I　P　A　L　G　F　G　T　Y　K　P　K　E   28

GTTCCCAAGAGTAAGTCACTGGAGGCTGCATGCCTAGCTCTAGATGTTGGTACCGCCATGTTGATACTGCTTATCGATACCAAGTAGAA   180
V　P　K　S　K　S　L　E　A　A　C　L　A　L　D　V　G　Y　R　H　V　D　T　A　Y　A　Y　Q　V　E   58

GAGGAGATAGGACAGGCCATTCAAAGCAAGATTAAAGCTGGGGTTGTAAAGAGAAGACCTGTTCGTCGTCACTACAAAGCTTTGGTCGACT   270
E　E　I　G　Q　A　I　Q　S　K　I　K　A　G　V　V　K　R　E　D　L　F　V　T　T　K　L　W　C　T   88

TGCTTTCGACCAGCTGCGTCAAGCCTGCCTTGGAAAAGTCACTGAAAAAGCTTCAGCTGGATTATCTTGATCTTTACATTATGCATTAC   360
C　F　R　P　E　L　V　K　P　A　L　E　K　S　L　K　K　L　Q　L　D　Y　V　V　D　L　Y　I　M　H　Y   118

CCAGTGCCAATGAAGTCAGGGGATAATGATTTTCCAGTAAATGAGCAAGGAAATCTCGTTGTTGGACACTGTGGATTTCGTTGTGACACATGG   450
P　V　P　M　K　S　G　D　N　D　F　P　V　N　E　Q　G　K　S　L　L　D　T　V　D　F　C　D　T　W   148

GAGAGGTTGGAGGAGTGTAAGGATGCAGGATTGGTCAAGTCCATTGGGGTGTCCAACTTAACCACAGGCAGCTGGAGCGAATCCTCAAT   540
E　R　L　E　E　C　K　D　A　G　L　V　K　S　I　G　V　S　N　F　N　H　R　Q　L　E　R　I　L　N   178

AAGCCCAGGACTGAAGTACAAACCTGTCTGCAACCAGCTTGAATGTCATCTCTATTTGAACCAGCGTAAGCTACTGGATTACTGCGAATCA   630
K　P　G　L　K　Y　K　P　V　C　N　Q　V　E　C　H　L　Y　L　N　Q　R　K　L　L　D　Y　C　E　S   208

AAAGACATTGTTCTTGTTGTTGCTTACGGTGCTCTCGGGGACCCAGCGATATAAAGAATGGGTTGACCAGAACTCCCCAGTTCTCTTGAATGAT   720
K　D　I　V　L　V　A　Y　G　A　L　G　T　Q　R　Y　K　E　W　V　D　Q　N　S　P　V　L　L　N　D   238

CCAGTTCTTTGTTGATTGTGGCCAAAAAGAACAACGCGAAGTCCTGCCTTGATTGCACTTCGATACCTGATTCAACGTGGGATTGTGCCCCTG   810

P V L C D V A K K N K R S P A L I A L R Y L I Q R G I V P L   268

GCCCAGAGTTTCAAAGACAGAATGAGATGAGAGAATTGCAGGTTTTTGGATTTCAGTTGTCCCTGAGGACATGAAAACACTAGATTGCC   900

A Q S F K E N E M R E N L Q V F G F Q L S P E D M K T L D G   298

CTGAACAAAACTTTCGATACCTTCCAGCAGAGTTCCTTGTTGACCCAGTATCCATTTGTGGAGGAATATTAACATGCGGACCTA   990

L N K N F R Y L P A E F L V D H P E Y P F V E E Y *   323

ATCATGCCTTCTGCCCTGATGTCCCGTGTGTGTGGACAGTGATGCTGCGAATATGACCAAGATGACGACTTGTTGGATCGGACTTGTCATTTCTGA   1080

TCAATCTTGGTTGCTTAGCAACTCACATTCAGCTGAAGCTTTAATTAATTGATCTCAAAGAAATGAATATAATTTTCATGATGCTTTGAA   1170

ATAAATATGAATTTTTCTCTTAAAAAAAAAAAA

Figure 1. Nucleotide and deduced amino acid sequences of the cDNA for
AKR1C22. The single letter symbol for each amino acid is placed below the
central nucleotide of its codon. The termination codon (TAA) is labeled with an
asterisk. A putative polyadenylation signal is underlined. The poly(A) sequence
is indicated with a broken line.

```
                10        20        30        40        50        60
AKR1C22  MSSKQHCVKL NDGHLIPALG FGTYKPKEVP KSKSLEAACL ALDVGYRHVD TAYAYQVEEE
AKR1C12  ---------- ---------- ---------- ---------- ------L--- ----------
AKR1C13  ------Y--- ---------- ---------- ---------- ---------- ----------
RAKB     -D-IS----- -H--F----- ------EK-- --------H- -I-A----I- ------I---
RAKD     ----H----- ----F----- ---SI-N--- -------VH- -I-A--H-I- --S---I---
M6D      XXXXXXXXXX X---C----- ------I--- ---AM---N- -IG-----I- ------I---
                70        80        90       100       110       120
AKR1C22  IGQAIQSKIK AGVVKREDLF VTTKLWCTCF RPELVKPALE KSLKKLQLDY VDLYIMHYPV
AKR1C12  ---------- ---------- I--------- ---------- ---------- ----------
AKR1C13  ---------- ---------- -------G-- ---------- ----S----- ----LI----
RAKB     ---------- --------M- I--------- ---------- ----N----- A---------
RAKD     ---------- --------M- I--------- ---------- ----N----- A---------
M6D      -------N-- --I-----M- I--------- Q----R-S-- -XXX----EH ---F------
               130       140       150       160       170       180
AKR1C22  PMKSGDNDFP VNEQGKSLLD TVDFCDTWER LEECKDAGLV KSIGVSNFNH RQLERILNKP
AKR1C12  ---------- ---------- ---------- ---------- ---------- ----------
AKR1C13  ---P---ES- LD-N--F--- ---------- ---------- ---------- --------N-
RAKB     ---------- -D-K------ --------M --K------- ---------- KG---L----
RAKD     ------KYL- -DDN--W--- --------M --K------- ---------- K----LI---
M6D      ---A------ LD----L--- -----A---A --KXX----- ---------M ----------
               190       200       210       220       230       240
AKR1C22  GLKYKPVCNQ VECHLYLNQR KLLDYCESKD IVLVAYGALG TQRYKEWVDQ NSPVLLNDPV
AKR1C12  ---------- ---------- ---------- ---------- ---------- ----------
AKR1C13  ---------- --------S ------K--- ---------- ---------- ----------
RAKB     ---------- --------S ------K--- ---------- ---------- ----------
RAKD     ---------- ------M--S ------K--- ---------- ---------- ------D---
M6D      ---------- ----V-N--S ------K--- -----F---- ---------- D---------
               250       260       270       280       290       300
AKR1C22  LCDVAKKNKR SPALIALRYL IQRGIVPLAQ SFKENEMREN LQVFGFQLSP EDMKTLDGLN
AKR1C12  ---------- ---------- ---------- ---------- ---------- ----------
AKR1C13  ------R--- ---------- F--------- ---------- ----E----- ----------
RAKB     ---------- T--------- V---V----- ---------- ----D----- ----------
RAKD     ---------- ---------- V--EV----- ---------- ----D---H- ----------
M6D      --GG-KXXX- ---------- V---V----- --Y-S--K-- ----E----- ----I-----
               310       320
AKR1C22  KNFRYLPAEF LVDHPEYPFV EEY
AKR1C12  ---------- ---------- ---
AKR1C13  ---------- -A-------S ---
RAKB     ------S--- -AG------S ---
RAKD     ------S--- -AG--GC--S ---
M6D      --------Q- FAD------S --
```

Figure 2. Sequence alignment of AKR1C22 with highly homologous proteins in the AKR superfamily. AKR1C12 and AKR1C13: mouse aldo-keto reductases, RAKB and RAKD: rat aldo-keto reductases (AKR1C16 and AKR1C17), M6D: hamster morhpine 6-dehydrogenase. Identical residues are enumerated with dashes and X represents the undetermined residue by peptide sequencing.

value (37,042) calculated from the deduced amino acid sequence and in good agreement with the value (35,000) of undenatured AKR1C22 estimated by gel filtration, indicating that AKR1C22 is a monomeric enzyme.

Characterization of His-tag fusion AKR1C22 and AKR1C22

The effect of pH on the dehydrogenase and reductase activities of recombinant His-tag fusion AKR1C22 and AKR1C22 was examined (data not shown). The enzymes exhibited similar NADH-dependent maximal activities around pH 5.5. The NAD^+-dependent maximal activities of them were observed between pH 9 and 10. The substrate specificity of the enzymes were then determined (Table I). Both enzymes showed nearly the same relative dehydrogenase activities for S-indan-1-ol, $trans$-benzene dihydrodiol and 2-cyclohexen-1-ol. The enzymes reduced aromatic aldehydes and dicarbonyl compounds including quinones, as well as diacetyl with comparable activities. Recombinant AKR1C22 also reduced isatin and acetoin. The enzyme more efficiently utilized dicarbonyl compounds as substrates than other carbonyl compounds and alicyclic alcohols. The results indicate that AKR1C22 is a dihydrodiol dehydrogenase and has typical catalytic properties of enzymes in the AKR superfamily. Comparison of the properties of the two enzymes show that although the His-tag fusion protein lacks some activity, it is easily purified making it useful to analyze structure-function relationships of AKR1C22.

Kinetic analysis of AKR1C22 showed that the K_m and k_{cat} values for benzene dihydrodiol were 2048 μM and 36 min^{-1}, respectively, and these gave the k_{cat}/K_m value of 18 min^{-1} mM^{-1}. The K_m values of the enzyme for NAD^+ and $NADP^+$ were 14 μM and 418 μM, respectively, in the assay at pH 7.4 with 1 mM S-tetralol as the fixed substrate and those for NADH and NADPH were 9.2 μM and 203 μM, respectively, in the assay at pH 7.4 with 2 mM diacetyl as the fixed substrate. The k_{cat}/K_m values for NAD^+ (1429 min^{-1} mM^{-1}) and NADH (1957 min^{-1} mM^{-1}) were higher than those for $NADP^+$ (36 min^{-1} mM^{-1}) and NADPH (330 min^{-1} mM^{-1}), respectively. The results demonstrate that the enzyme prefers NAD(H) to NADP(H) as the cofactor. The cofactor preference of AKR1C22 is observed in the case of hamster M-6-D (*31*) and AKR1C12 (*26*), and it is different from those of other enzymes in the AKR1C subfamily and DDs purified from mouse (*19-21*). Of the 14 residues involved in cofactor binding of AKR1C2 (*16*), 9 residues (Tyr24, Asp50, Ser166, Asn167, Tyr216, Leu219, Ser271, Glu279, and Asn280) were conserved in the AKR1C22 amino acid sequence, but Ser217, Ser221, Lys270, Tyr272, and Arg276 in the AKR1C2 amino acid sequence were replaced by Gly, Thr, Glu, Phe, and Glu, respectively. Crystal structure analyses of AKR1C9 (*15*) and AKR1C2 (*16*), and site-directed mutagenesis study on AKR1C4 (*32*) have suggested that 2 residues (Lys270 and Arg276) interact with the 2'-phosphate of the adenosine moiety of NADP(H).

Table I. Substrate specificity of His-tag fusion AKR1C22 (His/ AKR1C22) and AKR1C22

Substrate	Concentration (mM)	Relative activity (%) His/ AKR1C22	AKR1C22
Dehydrogenase activity			
S-Tetralol	1.0	100	100
R-Tetralol	1.0	0.3	0.3
S-Indan-1-ol	1.0	162	167
R-Indan-1-ol	1.0	nd	38
trans-Benzene dihydrodiol	1.0	40	41
2-Cyclohexen-1-ol	1.0	18	16
1-Acenaphthenol	0.5	48	61
Reductase activity			
2-Nitrobenzaldehyde	1.0	54	43
3-Nitrobenzaldehyde	1.0	36	39
Pyridine-2-aldehyde	1.0	101	102
Pyridine-3-aldehyde	1.0	17	20
Pyridine-4-aldehyde	1.0	37	38
2, 3-Pentanedione	1.0	214	212
2, 3-Hexanedione	1.0	131	109
1, 2-Cyclohexadione	1.0	177	115
Dimethy-2-oxoglutarate1.0	1.0	266	253
Acenaphthequinone	0.025	177	147
Camphorquinone	0.5	31	22
Diacetyl	1.0	266	220
Isatine	0.4	nd	177
Acetoin	1.0	nd	8

The dehydrogenase and reductase activities were measured at pH 7.4 with 0.25 mM NAD^+ and 0.1 mM NADH, respectively, as cofactors. Enzyme activity of each substrate is expressed relative to the dehydrogenase activity for S-tetralol of His-tag fusion mDD (167 nmol/min/mg protein) and mDD (286 nmol/min/mg protein). nd, not determined.

In AKR1C22 and hamster M-6-D the two residues that contribute to NAD(H) preference are supposed to be Glu270 and Glu276, however, in AKR1C13 these residues are present yet NADP(H) is the preferred cofactor (26).

Inhibitor Sensitivity of Recombinant mDD

The dehydrogenase and reductase activities of recombinant AKR1C22 were inhibited by hydroxysteroid dehydrogenase inihibitors such as hexestrol, medroxyprogesterone, ethacrynic acid and indomethacin (Table II). Lithocholic acid showed relatively low inhibitory effects on both activities of the enzyme, while steroidal hormones tested exhibited higher inhibition on the reductase activity than on the dehydrogenase activity. The enzymatic activity of AKR1C22 may be regulated physiologically by such endogenous compounds. Lithocholic acid is an inhibitor of AKR1C2, M-6-D's from hamster (*31*) and rabbit (*33*), and AKR1C22. The amino acid sequences of AKR1C22 and M-6-D's of hamster and rabbit show that of the 9 amino acids implicated in the binding of ursodeoxycholate to AKR1C2 (*16*), three have been substituted. Thus, Val128, Ile129, and Leu308 have been replaced by Asp, Phe and Ala, respectively. Thus, the remaining six conserved residues may contribute to the interaction of lithcholic acid with the enzymes.

Table II. Effects of Inhibitors on AKR1C22

Inhibitor	Concentration (mM)	Inhibition (%) Oxidation	Reduction
Hexestrol	0.01	16	48
Medroxyprogesterone	0.01	10	30
Quercitrin	0.01	13	13
Lithocholic acid	0.05	71	82
Taurucholic acid	0.05	9	13
Ursodeoxycholic acid	0.05	5	12
Progesterone	0.05	4	49
Androsterone	0.05	13	47
Testosterone	0.05	12	33
Ethacrynic acid	0.1	74	76
Indomethacin	0.1	41	21
Sorbinil	0.1	14	14
Dicoumarol	0.1	8	3
Benzoic acid	1.0	14	11
Pyrazole	1.0	5	9

The dehydrogenase and reductase activities were measured at pH 7.4 with 1 mM S-tetralol and 0.25 mM NAD^+, and 2 mM diacetyl and 0.1 mM NADH, respectively.

Expression of mRNA for AKR1C22 in Mouse Tissues

The expression of mRNAs for AKR1C22 and AKR1C6 were analyzed by RT-PCR using primers specific for each mRNA. As shown in Figure 3, the expected 352 bp and 740 bp bands for AKR1C22 and AKR1C6, respectively, were specifically amplified from total RNAs of various tissues. mRNA for AKR1C22 was ubiquitously expressed in all tissues, while mRNA for AKR1C6

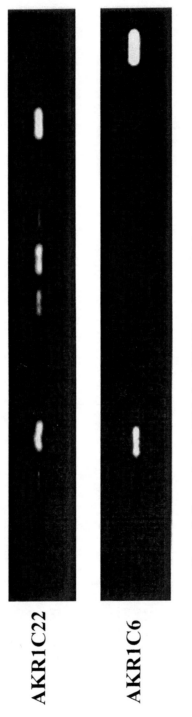

Figure 3. Expression of mRNAs for AKR1C22 and mouse 17β-HSD (AKR1C6). The samples from brain (A), heart (B), lung (C), liver (D), kidney (E), spleen (F), testis (G), small intestine (H), stomach (I) and blood (J) were analyzed. Plasmids containig cDNA for AKR1C22 (K) and AKR1C6 (L) were used as control templates.

was strictly expressed in the liver. The most significant amplifications of the cDNA fragment was observed in the samples of the liver and small intestine. Moderate amplifications were detected in samples of the testes, stomach and lung.

Conclusions

Mouse DD (AKR1C22) cloned in the present study is a new member of the aldo-keto reductade superfamily and may be involved in the detoxification of xenobiotics, in the metabolism of some endogenous carbonyl compounds and partly in the activation of carcinogenic compounds derived from polycyclic aromatic hydrocarbons. The enzymatic activity of AKR1C22 may be regulated by endogenous compounds such as lithocholic acid and some steroidal hormones.

Acknowledgement

This work was supported in part by grant-in-aid for scientific research (C) KAKENHI (13672291) from Japan Society for the Promotion of Science (JSPS).

References

1. Glatt, H. R.; Cooper, C. S.; Grobver, P. L.; Sims, P.; Bentley, P.; Merdes, M.; Waechter, F.; Vogel, K.; Guenther, T. M.; Oesch, F. *Science* **1982,** 215, 1507-1509.
2. Smithgall, T. E.; Harvey, R. G.; Penning, T. M. *J. Biol. Chem.* **1986,** 261, 6184-6191.
3. Murty, V. S.; Penning, T. M. *Chem.-Biol. Interact.* **1992,** 84, 169-188.
4. Shou, M.; Harvey, R. G.; Penning, T. M. *Carcinogenesis* **1993,** 14, 475-482.
5. McCoull, K. D.; Rindgen, D.; Blair, I. A.; Penning, T.M. *Chem. Res. Toxicol.* **1999,** 12, 237-246.
6. Hara, A.; Taniguchi, H.; Nakayama, T.; Sawada, H. *J. Biochem. (Tokyo)* **1990,** 108, 250-254.
7. Deyashiki, Y.; Taniguchi, H.; Amano, T.; Hara, A.; Sawada, H. *Biochem. J.* **1992,** 282, 741-746.
8. Ohara, H.; Nakayama, T.; Deyashiki, Y.; Hara, A.; Tukada, F. *Biochim. Biophys. Acta* **1994,** 1215, 59-69.
9. Hara, A.; Matsuura, K.; Tamada, Y.; Sato, K.; Miyabe, Y.; Deyashiki, Y.; Ishida, N. *Biochem. J.* **1996,** 313, 373-376.
10. Jez, J. M.; Flynn, G.; Penning, T. M. *Biochem. Pharmacol.* **1997,** 54, 639-647.

11. Jez, J. M.; Bennet, M. J.; Schlegel, B. P.; Lewis, M.; Penning, T. M. *Biochem. J.* **1997,** 326, 625-636.
12. Deyashiki, Y.; Ogasawara, A.; Nakayama, T.; Nakanishi, M.; Miyabe, Y.; Hara, A. *Biochem. J.* **1994,** 299, 545-552.
13. Shiraishi, H.; Ishikura, S.; Matsuura, K.; Deyashiki, Y.; Ninomiya, M.; Sakai, S.; Hara, A. *Biochem. J.* **1998,** 334, 399-405.
14. Matsuura, K.; Shiraishi, H.; Hara, A. ; Sato, K.; Deyashiki, Y.; Ninomiya, M.; Sakai, S. *J. Biochem. (Tokyo)* **1998,** 124, 940-946.
15. Bennet, M. J.; Albert, R. H.; Jez, J. M.; Penning, T. M.; Lewis, M. *Structure* **1997,** 5, 799-812.
16. Jin, Y.; Stayrook, S. E.; Albert, R. H.; Palackal, N. T.; Penning, T. M.; Lewis, M. *Biochemistry* **2001,** 40, 10161-10168.
17. Burczynski, M. E.; Harvey, R. G.; Penning, T. M. *Biochemistry* **1998,** 37, 6781-6790.
18. Palackal, N. T.; Lee, S. H.; Harvey, R. G.; Blair, I. A.; Penning, T. M. *J. Biol. Chem.* **2002,** 277, 24799-24808.
19. Bolcsak, L. E.; Nerland, D. E. *J. Biol. Chem.* **1983,** 258, 7252-7255.
20. Sawada, H.; Hara, A.; Nakayama, T.; Nakagawa, M.; Inoue, Y.; Hasebe, K.; Zhang, Y.-P. *Biochem. Pharmacol.* **1988,** 37, 453-458.
21. Nakagawa, M.; Tsukada, F.; Nakayama, T.; Matsuura, K.; Hara, A.; Sawada, H. *J. Biochem. (Tokyo)* **1989,** 106, 633-638.
22. Lo, Y.-M. D.; Mehal, W. Z.; Fleming, K. A. In *PCR Protocols;* Innis, M. A.; Gelfand, D. M.; Sninsky, J. J.; White, T. J., Eds.; Academic Press: New York, 1990; pp 113-118.
23. Bradford, M. M. *Anal. Biochem.* **1976,** 72, 248-254.
24. Platt, K.; Oesch, F. *Synthesis* **1977,** 7, 449-450.
25. Deyashiki, Y.; Ohshima, K.; Nakanishi, M.; Sato, K.; Matsuura, K.; Hara, A. *J. Biol. Chem.* **1995,** 270, 10461-10467.
26. Ikeda, S.; Okuda-Ashitaka, E.; Masu, Y.; Suzuki, T.; Watanabe, K.; Nakao, M.; Shingu, K.; Ito, S. *FEBS Lett.* **1999,** 459, 433-437.
27. Du, Y.; Tsai, S.; Keller, J. R.; Williams, S. C. *J. Biol. Chem.* **2000,** 275, 6724-6732.
28. Qin, K.; Cheng, K.-C. *Biochemistry* **1994,** 33, 3223-3228.
29. Kume, T.; Iwasa, H.; Shiraishi, H, Yokoi, T.; Nagashima, K.; Otsuka, M.; Terada, T.; Takagi, T.; Hara, A.; Kamataki, T. *Pharmacogenetics* **1999,** 9, 763-771.
30. Pawlowski, J. E.; Huizinga, M.; Penning, T. M. *J. Biol. Chem.* **1991,** 266, 8820-8825.
31. Todaka, T.; Yamano, S.; Toki, S. *Arch. Biochem. Biophys.* **2000,** 374, 189-197.
32. Matsuura, K.; Tamada, Y.; Sato, K.; Iwasa, H.; Miwa, G.; Deyashiki, Y.; Hara, A. *Biochem. J.* **1997,** 322, 89-93.
33. Yamano, S.; Ito, K.; Ogata, S.; Toki, S. *Arch. Biochem. Biophys.* **1997,** 341, 81-88.

Chapter 8

Efficient Syntheses of the Active Metabolites of Carcinogenic Polycyclic Aromatic Hydrocarbons

Ronald G. Harvey[1], Qing Dai[1], Chongzhao Ran[1], and Trevor M. Penning[2]

[1]The Ben May Institute for Cancer Research, University of Chicago, Chicago, IL 60637
[2]Department of Pharmacology, School of Medicine, University of Pennsylvania, Philadelphia, PA 19104

Carcinogenic polycyclic aromatic hydrocarbons (PAHs) require metabolic activation to exert their deleterious effects. Three pathways of activation have been proposed. The first entails initial formation of *trans*-dihydrodiol metabolites that undergo conversion to highly mutagenic diol epoxides that interact with DNA to form covalent adducts. The second path involves formation of PAH radical cations that react with DNA to form unstable depurinating adducts. In the third path, the *trans*-dihydrodiols are converted by aldo-keto reductase enzymes to reactive and redox-active *o*-quinones capable of interacting with DNA to form both stable and depurinating adducts. Investigations of the relative importance of these mechanisms require methods for the synthesis of the active metabolites involved. We now report efficient new syntheses of the *o*-quinones of benzo[*a*]pyrene (BPQ), 7,12-dimethyl-benz[*a*]anthracene (DMBAQ), and benz[*a*]anthracene (BAQ), convenient synthetic precursors of the corresponding *trans*-dihydrodiols, diol epoxides, and catechols.

115

Polycylic aromatic hydrocarbons (PAHs) are widespread environmental pollutants formed in the incomplete combustion of organic matter, and some PAHs are potent carcinogens (*1-4*). Prior investigations have shown that PAHs, such as benzo[*a*]pyrene (BP), are activated by the combined action of CYP1A1 and epoxide hydrolase enzymes to *anti-* and *syn-*diol epoxides that are highly mutagenic and tumorigenic (Figure 1) (*2,3*). In the case of BP, the *trans*-BP-7,8-diol yields predominantly (+)-*anti*-BPDE which has been shown to react with mutational hot spots in DNA to form stable adducts (*5,6*). Two additional pathways have also been proposed: (i) activation of PAHs by CYP peroxidase to form PAH radical cations that react with DNA to form unstable depurinating adducts (*7*), and (ii) oxidation of the *trans*-dihyrodiol metabolites by aldo-keto reductase enzymes (AKRs) to form catechols which enter into a redox cycle with the corresponding *o*-quinones (*8-11*). The *o*-quinones can form both stable and depurinating adducts with DNA (*12,13*), and redox cycling of the *o*-quinones generates reactive oxygen species (ROS) that can interact with DNA and lead to mutation.

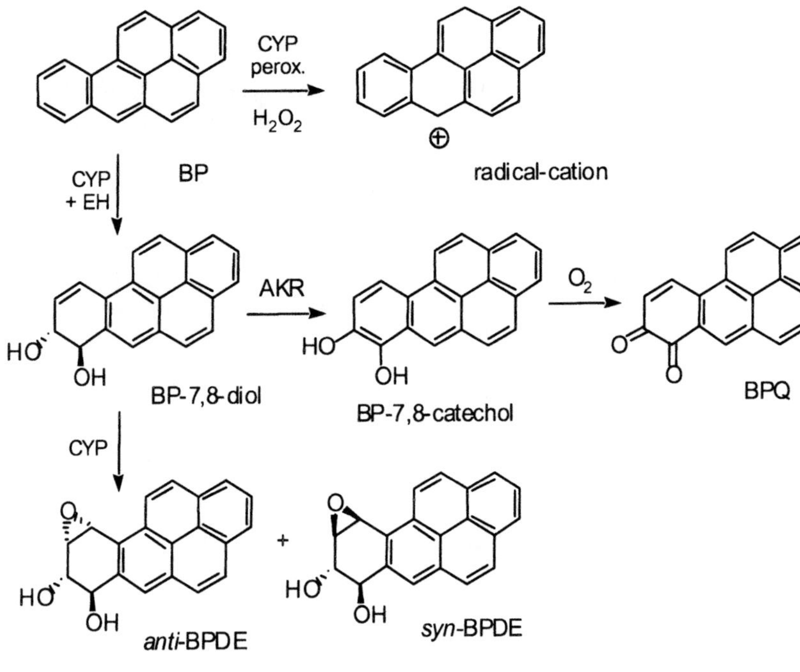

Figure 1. Pathways for activation of PAH carcinogens

In connection with studies to determine the relative importance of these pathways for carcinogenesis, we required methods for the convenient synthesis of the various activated metabolites involved (*trans*-dihydrodiols, diol epoxides, catechols, and quinones) and their adducts with deoxyguanosine (dG) and deoxyadenosine (dA). The PAH carcinogens selected for study were benzo[*a*]pyrene, benz[*a*]anthracene (BA), and 7,12-dimethylbenz[*a*]anthracene (DMBA). Although methods for the synthesis of the active metabolites of these PAHs have been described (*14-19*), they entail relatively complex multistep procedures that are best suited for small-scale preparations. More efficient synthetic methods were needed to make these compounds available in larger quantities for use as starting compounds for preparative-scale synthesis of the dG and dA adducts, as well as for biological studies. Although syntheses of the dG and dA adducts of *anti*- and *syn*-BPDE have been described (*20-25*), methods for preparation of the corresponding adducts of BA and DMBA have not been reported, nor have syntheses of the stable adducts of BPQ and other PAH quinones with dG and dA [see subsequent chapter for more recent findings]. We now report new, more efficient syntheses of the active metabolites of BP, BA, and DMBA that entail fewer synthetic steps than older procedures and are readily adaptable to preparations on any scale.

Results and Discussion

The quinone metabolites were chosen as the primary synthetic targets because they are known to be relatively stable compounds that may be readily converted to the corresponding *trans*-dihydrodiols, *anti*- or *syn*-diol epoxides, or catechols, as needed for subsequent studies.

Benzo[*a*]pyrene series

The synthetic approach to benzo[*a*]pyrene-7,8-dione (BPQ) is outlined in Figure 2. It entails in the initial step Suzuki coupling of 6-methoxy-naphthaleneboric acid with 2-bromo-1,3-dibenzaldehyde (1) catalyzed by $(Ph_3P)_4Pd$. It is based on the new method for the synthesis of PAHs recently reported (*26*). The dialdehyde 1 was synthesized from 2-bromoxylene (Figure 3) via photo-bromination with NBS (*27*) followed by hydrolysis of the product, 1,3-bis(dibromomethyl)-2-bromobenzene, with either a catalytic amount of $AgNO_3$ in CH_3CN or with aqueous formic acid (*28*). Both of these procedures were superior to the method of Mataka, et al. (*29*) [NaOAc/CaCO$_3$/Bu$_4$NBr] which required longer reaction time and afforded BPQ in consiserably lower yield.

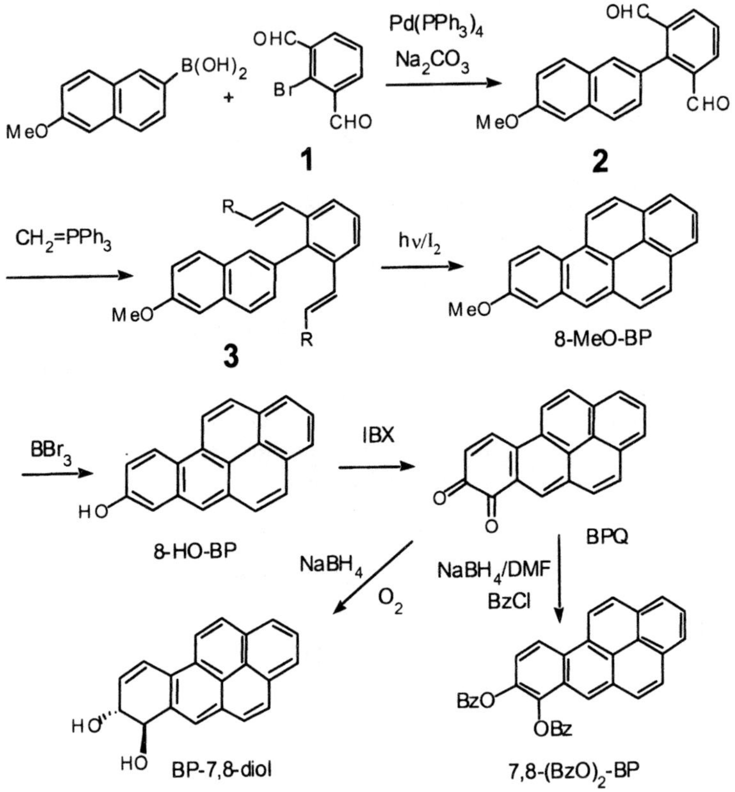

Figure 2. Synthesis of the active metabolites of benzo[a]pyrene

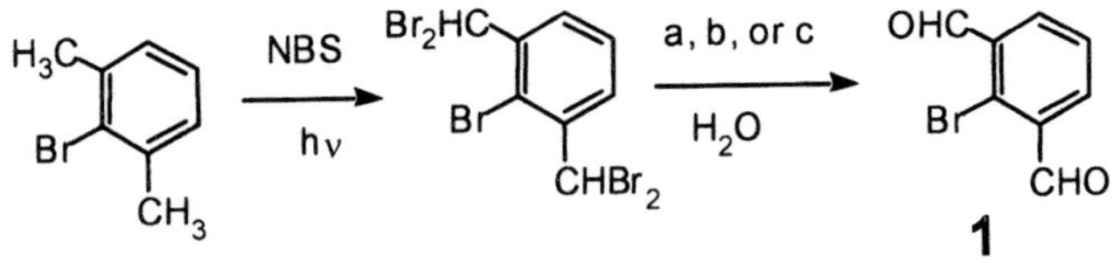

a = NaOAc/CaCO$_3$/Bu$_4$NHBr; b = AgNO$_3$/CH$_2$CN; c = formic acid.

Figure 3. Synthesis of 2-bromobenzene-1,3-dialdehyde

The dialdehyde product (2) obtained from coupling 6-methoxynaphthaleneboric acid with 1 entered into double Wittig reaction with methylenetriphenylphosphine to furnish the corresponding diolefin 3 (Figure 3). Oxidative photocyclization of 3 (R = H) with iodine and 1,2-epoxybutane by the usual procedure (26) provided 8-MeO-BP in moderate yield (45%). For this reason and because the photoreactions require dilute solutions that severely limit their scale, alternative methods of cyclization of 3 were explored. Acid-catalyzed cyclization of the methoxy analog of 3 (R = OMe) afforded 8-MeO-BP in considerably higher yield (95%). Demethylation of 8-MeO-BP with BBr$_3$ gave 8-HO-BP in good overall yield. The NMR spectra and physical properties of 8-MeO-BP and 8-HO-BP were in good agreement with those of the authentic compounds.

Conversion of PAH β-phenols to o-quinones is usually accomplished by oxidation with Fremy's reagent [(KSO$_3$)$_2$NO] (2). However, this reagent affords erratic results and requires aqueous conditions that are not compatible with the PAH reactants and quinone products. Accordingly, we investigated the use of o-iodoxybenzoic acid (IBX) as a potential alternative reagent. IBX is a mild oxidant that is widely employed for oxidation of alcohols (30). Reaction of 8-HO-BP with IBX took place smoothly in DMF to furnish the quinone BPQ essentially quantitatively. While this work was in progress, we came across a report on the regioselective oxidation of phenols with IBX to yield o-quinones (31). The reaction was found to be general for phenols containing at least one electron-donating group. Although 8-HO-BP lacks an electron-donating group, it reacts readily with IBX.

Reduction of BPQ with NaBH$_4$ in DMF by the procedure previously reported (32) gave the corresponding catechol [7,8-(HO)$_2$-BP] isolated as its dibenzoate or diacetate esters. The catechol itself is air-sensitive and undergoes visible transformation to the purple quinone on exposure to the atmosphere, whereas the dibenzoate or diacetate esters are stable in air. The auto-oxidizability of 7,8-[HO]$_2$-BP is consistent with its demonstrated facile conversion to BPQ in vivo (8-11).

Transformation of PAH o-quinones, such as BPQ, to the related trans-dihydrodiols, which in turn may be employed as starting compounds for preparation of the corresponding anti- and syn-diol epoxides, is well documented in prior studies (2,33). Thus, reduction of BPQ with NaBH$_4$ in the presence of O$_2$ gave trans-BP-7,8-diol (34), the synthetic precursor of the corresponding anti- and syn-BPDEs (2,33).

The syntheses of BPQ and other active metabolites of BP [7,8-(HO)$_2$-BP, trans-BP-7,8-diol, and the anti- and syn-BPDEs] reported make all of these compounds now readily available on any scale required for biological studies and as starting compounds for the synthesis of the corresponding DNA adducts.

7,12-Dimethylbenz[a]anthracene series

Synthesis of 7,12-dimethylbenz[a]anthracene-3,4-dione (DMBAQ) is outlined in Figure 4. It entails Suzuki coupling of 1,4-dimethyl-2-naphthalene boronic acid (4) with 2-bromo-5-methoxybenzaldehyde (5) catalyzed by (Ph₃P)₄Pd, and is based on the general method previously reported for the synthesis of PAH compounds (26). The boronic acid starting compound 4 was itself prepared from 1,4-dimethylnaphthalene (Figure 5) by Fe-catalyzed bromination (35) followed by conversion of the resulting 2-bromo-1,4-dimethyl-naphthalene to the corresponding Lithium compound, reaction of the latter with

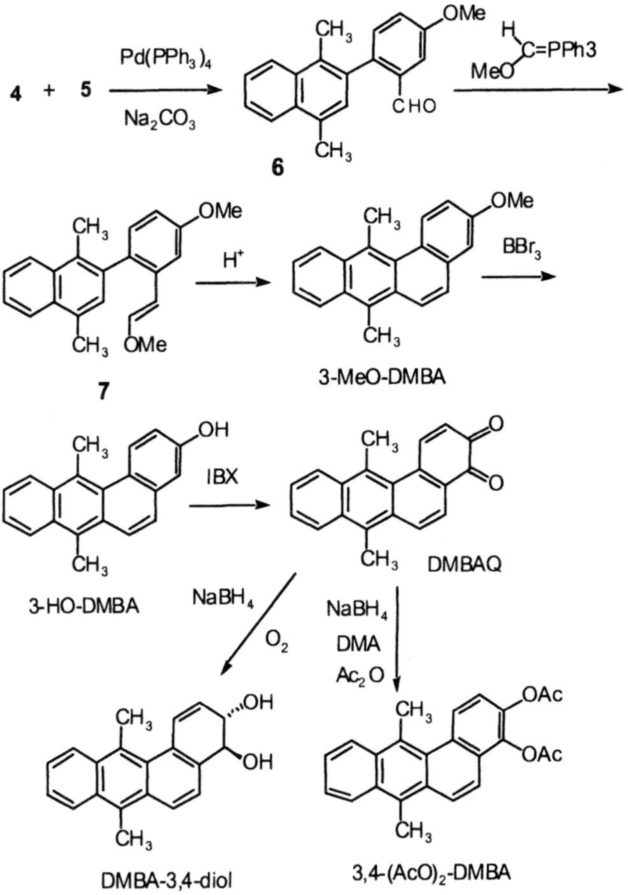

Figure 4. Synthesis of DMBAQ and other active metabolites of DMBA

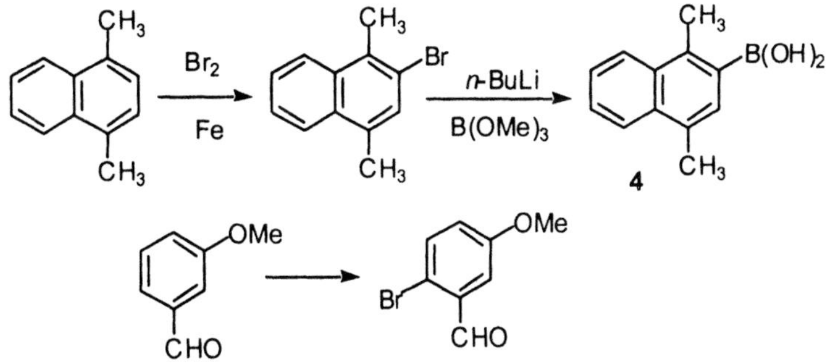

Figure 5. Synthetic precursors of DMBAQ

trimethylborate, and hydrolysis (36). The aldehyde **5** was prepared from 3-methoxybenzaldehyde via bromination by the published procedure (37-39).

Suzuki coupling of **4** with **5** in the presence of $(Ph_3P)_4Pd$ took place smoothly to furnish the product **6**. Wittig reaction of **6** with methoxymethylene-triphenylphosphine generated *in situ* from (methoxymethyl) triphenylphosphonium chloride and phenyllithium at −65 °C in ether afforded the enol ether **7** as a mixture of *E* and *Z* isomers (by NMR analysis). Acid-catalyzed cyclization of **7** by treatment with methanesulfonic acid provided 3-methoxy-7,12-dimethylbenz[*a*]anthracene (3-MeO-DMBA). Removal of the methyl group with BBr_3 provided the phenol, 3-HO-DMBA, in good overall yield. The proton and carbon NMR spectra and physical properties of 3-HO-DMBA were in good agreement with those reported (19).

Conversion of 3-HO-DMBA to the related *o*-quinone (DMBAQ) and the corresponding catechol [3,4-$(HO)_2$-DMBA] were accomplished by the methods employed for the synthesis of the related quinone and catechol metabolites of BP. Thus, oxidation of 3-HO-DMBA with IBX in DMF took place smoothly to furnish DMBAQ essentially quantitatively as a deep purple crystalline solid. The proton NMR spectrum and physical properties of DMBAQ were in good agreement with those reported previously (19,35,40,41). Reduction of DMBAQ with $NaBH_4$ in DMF (32) provided 3,4-$(HO)_2$-DMBA (41) isolated as its dibenzoate ester.

Transformation of DMBAQ to the related *trans*-dihydrodiol (DMBA-3,4-diol) via reduction with $NaBH_4/O_2$ (19) or $LiAlH_4$ (40), and the further transformation of DMBA-3,4-diol to the DMBA *anti*- and *syn*-diol epoxides was previously described (19,40).

Benz[a]anthracene series

Synthesis of benz[a]anthracene-3,4-dione (BAQ) by direct application of the method employed for preparation of DMBAQ (Figure 4) was not feasible because cyclization of the unmethylated analog of **7** was anticipated to take place to the α-position, resulting in formation of a chrysene rather than a benz[a]anthracene derivative. In order to block this mode of cyclization, methoxy groups were introduced into the 1,4-positions of the naphthalene ring. Although only a single blocking group was needed, it was more convenient to employ the dimethoxy analog.

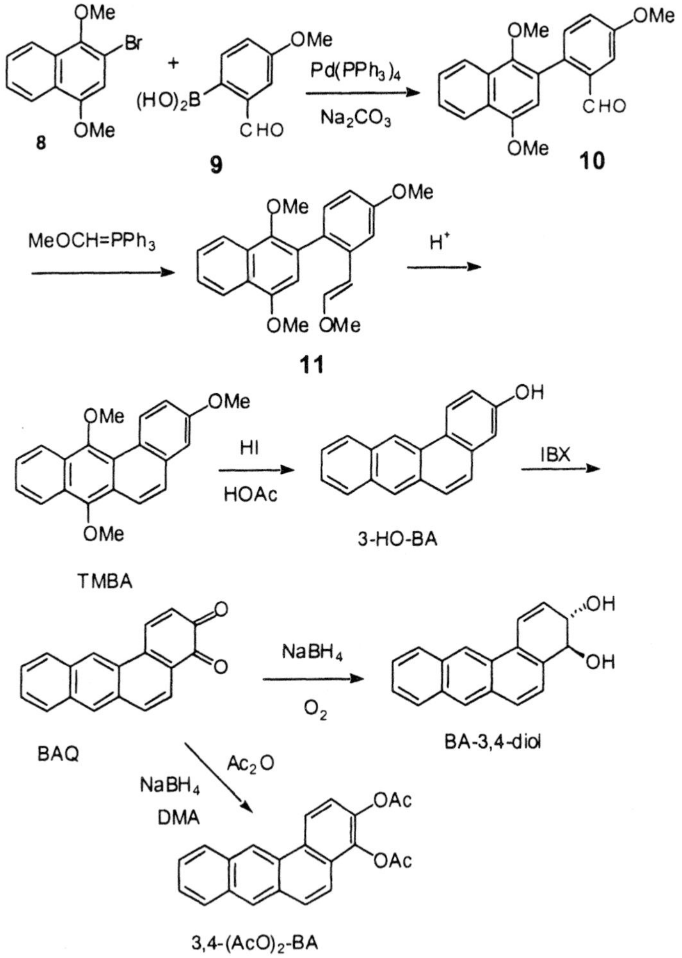

Figure 6. Synthesis of BAQ and other active metabolites of BA

The synthetic route to BAQ entails in the key step Suzuki coupling of 2-bromo-1,4-dimethoxy-2-naphthalene (8) with 2-formyl-4-methoxyboronic acid (9) catalyzed by $(Ph_3P)_4Pd$ (Figure 6).

The starting aryl bromide 8 was purchased from the Aldrich-Sigma Co., and the 2-formyl-4-methoxyboronic acid (9) was prepared from 2-bromo-4-methoxybenzaldehyde by the published method (39). Suzuki reaction of 8 with 9 afforded smoothly the coupled product 10. Reaction of 10 with methoxymethylenetriphenylphosphine generated *in situ* from (methoxymethyl)triphenyl-phosphonium chloride and *n*-butyllithium at –65 °C in ether furnished the enol ether 11. Methanesulfonic acid-catalyzed cyclization of 11 gave 3,7,12-trimethoxybenz[*a*]anthracene (TMBA). Selective removal of the 7,12-methoxy groups of TMBA was accomplished by reduction with hydriodic acid in acetic acid by the method of Konieczy and Harvey (42). The regioselectivity of this reaction is ascribed to the greater reactivity of the *meso* region positions relative to other positions of BA for all types of reactions. The remaining methoxy group underwent concurrent demethylation to furnish directly the corresponding phenol, 3-HO-BA. The spectral and physical properties of 3-HO-BA were in good agreement with those reported (43). Oxidation of 3-HO-BA with IBX in DMF furnished BAQ in good overall yield.

Reduction of BAQ with $NaBH_4$ in DMA by the usual procedure (32) provided the corresponding catechol [3,4-$(HO)_2$-BA] isolated as its diacetate [3,4-$(AcO)_2$-BA]. Reduction of BAQ with $NaBH_4/O_2$ was previously shown to take place stereospecifically to yield *trans*-3,4-dihydroxybenz[*a*]anthracene (BA-3,4-diol), and its further transformation to the corresponding *anti*- and *syn*-BA-diol epoxides was also previously described (19,40).

Discussion

The novel syntheses of BPQ, DMBAQ, and BAQ reported herein make these compounds and the other related oxidized metabolites of the parent PAHs now readily available for a wide range of chemical and biological studies. The prior relative inaccessibility of these compounds to many investigators has seriously retarded investigations of the mechanisms of PAH carcinogenesis at the molecular-genetic level.

The synthetic route to BPQ outlined in Figure 2 entails considerably fewer steps than the prior methods of synthesis (2,3), and provides BPQ in superior overall yield. In addition, it employs relatively inexpensive, nonhazardous reagents, involves relatively straightforward procedures that present minimal hazard to personnel, and is suitable for preparation on virtually any scale. For small scale preparations (<500 mg), it may be convenient to cyclize the divinyl intermediate 3 (R = H) photochemically to obtain the phenol ether 4, but for larger scale preparations cyclization is most satisfactorily accomplished by acid-catalyzed cyclization of the dimethoxyvinyl compound 3 (R = OMe). The photochemical method requires relatively dilute solutions and large volumes of solvents (44), whereas this is not the case for acid-catalyzed cyclization. In our experience, the yields are generally higher by the acid-catalyzed method.

Another important advantage of this new synthesis of BPQ is that it also provides more convenient access to all of the oxidized metabolites of BP suspected to play a role in its mechanism of carcinogenesis, as well. Thus, reduction of BPQ with $NaBH_4$ in DMF provides the corresponding catechol $(7,8-[HO]_2-BP)$ isolated as the stable diacetate or dibenzoate ester. Reduction of BPQ with $NaBH_4$ in EtOH in the presence of O_2 furnishes the corresponding *trans*-dihydrodiol (*trans*-BP-7,8-diol), and this dihydrodiol is the immediate synthetic precursor of the highly mutagenic *anti*- and *syn*-diol epoxides (*anti*- and *syn*- BPDE) (2,33).

The significant features of this synthetic methodology are the use of Suzuki coupling combined with either photochemical or acid-catalyzed cyclization to generate a PAH phenol with the appropriate PAH skeleton followed by oxidation of the phenolic intermediate with IBX to yield the PAH quinone.

Application of this general approach to the syntheses of analogous quinones of DMBA and BA was equally successful. The new syntheses of DMBAQ and BAQ reported (Figures 4 and 6) represent substantial improvements over the prior reported methods of preparation of DMBAQ (19,35,40,41) and BAQ (34,40,45). Reduction of DMBAQ and BAQ with $NaBH_4$ in DMA by the method previously reported (32) furnished the corresponding catechols [3,4-$(HO)_2$-DMBA and 3,4-$(HO)_2$-BA] as their diacetate or dibenzoate derivatives in good yield. Since DMBAQ and BAQ are the established synthetic precursors of the corresponding 3,4-dihydrodiols and *anti*- and *syn*-diol epoxide derivatives, these compounds are also now conveniently accessible via this route.

Acknowledgement

This investigation was supported by grant PO1 CA92537 from the National Cancer Institute.

References

1. International Agency for Research on Cancer. Monographs on the Evaluation of the Carcinogenic Risk of Chemicals to Humans. *Polynuclear Aromatic Compounds, Part 1, Chemical, Environmental and Experimental Data,* Vol. 32, IARC, Lyon, France, 1983.
2. Harvey, R. G. *Polycyclic Aromatic Hydrocarbons: Chemistry and Carcinogenesis,* Cambridge University Press, Cambridge, U.K., 1991.

3. Harvey, R. G., *Polycyclic Hydrocarbons and Carcinogenesis;* American Chemical Society: Washington, DC, 1985.
4. Harvey, R. G. *The Handbook of Environmental Chemistry*, Volume 3, Part I: *PAHs and Related Compounds*, O. Hutzinger, Editor-in-Chief, A. Neilson, Volume Ed., Chap. 1, Springer, Berlin, Heidelberg, 1997, pp 1-54.
5. Dipple, A. *DNA Adducts: Identification and Biological Significance*, Hemminki, K.; Dipple, A.; Segerbäck, D.; Kadulbar, F. F.; Shuker, D.; Bartsch, H., Eds., IARC Scientific Publication No. 125, 1994, pp. 107-129.
6. Harvey, R. G.; Geacintov, N. E. *Acc. Chem. Res.* **1988**, *21*, 66-73.
7. Cavalieri, E. L.; Rogan, E.G. *Xenobiotica*, **1995**, *25*, 677-688.
8. Smithgall, T. E.; Harvey, R. G.; Penning, T. M. *J. Biol. Chem.* **1986**, *261*, 6184-6191.
9. Smithgall, T. E.; Harvey, R. G.; Penning, T. M. *Cancer Res.* **1988**, *48*, 1227-1232.
10. Smithgall, T. E.; Harvey, R. G.; Penning, T.M. *J. Biol. Chem.* **1988**, *263*, 1814-1820.
11. Penning, T. M.; Ohnishi, S. T.; Ohnishi, T. Harvey, R. G. *Chem. Res. Toxicol.* **1996**, *9*, 84-92.
12. Shou, M.; Harvey, R. G.; Penning, T. M. *Carcinogenesis* **1993**, *14*, 475.
13. McCoull, K. D.; Rindgen, D.; Blair, I. A.; Penning, T. M. *Chem. Res. Toxicol.* **1999**, *12*, 237-246.
14. Harvey, R. G. *Polycyclic Aromatic Hydrocarbons: Chemistry and Carcinogenesis,* Cambridge University Press, Cambridge, U.K., 1991, Chap. 14, pp 330-359.
15. Beland, F. A; Harvey, R. G. *J. Chem. Soc. Chem. Commun.* **1976**, 84-85.
16. Yagi, H.; Thakker, D. R.; Hernandez, O.; Koreeda, M.; Jerina, D. M. *J. Am. Chem. Soc.* 99, **1977**, 1604-1611.
17. Harvey, R. G.; Sukumaran, K. *Tetrahedron Lett.* **1977**, 28, 2387-2390.
18. Lee, H.; Harvey, R. G.; Cortez, C.; Kiselyov, A. *Bioorg. Med. Chem. Lett.* **1997**, *7*, 443-446.
19. Lee, H.; Harvey, R. G. *J. Org. Chem.* **1986**, *51*, 3502-3507.
20. Lee, H.; Hinz, M.; Stezowski, J. J.; Harvey, R. G. *Tetrahedron Lett.* **1990**, *31*, 6773-6776.
21. Lee, H.; Luna, E.; Hinz, M.; Stezowski, J. J.; Kiselyov, A. S.; Harvey, R. G. *J. Org. Chem.* **1995**, *60*, 5604-5613.
22. Laksman, M. K.; Sayer, J. M.; Jerina, D. M. *J. Am. Chem. Soc.* **1991**, *113*, 6589-6594.
23. Laksman, M. K.; Sayer, J. M.; Jerina, D. M. *J. Org. Chem.* **1992**, *57*, 3438-3443.
24. Cooper, M. D.; Hodge, R. P.; Tamura, P. J.; Wilkinson, A. S.; Harris, C. M.; Harris, T.M. *Tetrahedron Lett.* **2000**, *41*, 3555-3558.
25. Cosman, M.; Ibanez, V.; Geacintov, N. E.; Harvey, R. G. *Carcinogenesis* **1990**, *11*, 1667-1672.

26. F.-J. Zhang, F.-J.; Cortez, C.; Harvey, R. G. *J. Org. Chem.* **2000**, *65*, 3952.
27. Mataka, S.; Liu, G. B.; Sawada, T.; Kurisa, M.; Tashiro, M. *Bull. Chem. Soc. Japan* **1994**, *67*, 1113-1119.
28. Makasza,, M.; Owczarczyk, Z. *J. Org. Chem.* **1989**, *54*, 5094-5100.
29. Mataka, S.; Liu, G.-B.; Sawada, T.; Tori-I, A.; Tashiro, M., *J. Chem. Res. (S)* **1995**, 410-411.
30. Nicolaou, K. C.; Momtagnon, T.; Baran, P. S.; Zhong, Y.-L. *J. Am. Chem. Soc.* **2002**, *124*, 2245-2258.
31. Magdziak, D.; Rodriguez, A. A.; Van De Water, R. W.; Pettus, T. R. R. *Org. Lett.* **2001**, *4*, 285-288.
32. Cho, H.; Harvey, R.G. *J.C.S Perkin I.* **1976**, 836-839.
33. Harvey, R. G.; Fu, P. P. In, *Polycyclic Hydrocarbons and Cancer: Environment, Chemistry, and Metabolism.* Vol. 1. Gelboin, H.V.; T'so, P.O.P., Eds., Academic Press, New York, 1978, pp. 133-165.
34. Platt, K. L.; Oesch, F. *J. Org. Chem.* **1983**, *48*, 265-268.
35. Sharma, P. K. *Syn. Commun.* **1993**, *23*, 389-394.
36. Washburn, R. M.; Levens, E.; Albright, C. F.; Billig, F. A. *Org. Syn.*, Coll. Vol. 4, 1963, p. 68-72.
37. Harmata, M.; Kahraman, M. *J. Org. Chem.* **1999**, *64*, 4949-4952.
38. Léo, P.-M.; Morin, C.; Philouze, C. *Org. Lett.* **2002**, *4*, 2711-2714.
39. Kumar, S. *J. Chem. Soc. Perkin Trans. 1* **1998**, 3157-3161.
40. Sukumaran, K. B.; Harvey, R. G. *J. Org. Chem.* **1980**, *45*, 4407-4413.
41. Newman, M. S.; Khanna, J. M.; Khanna, V. K.; Kanakarajan, K. *J. Org. Chem.* **1979**, *44*, 4994-4995.
42. Konieczny, M.; Harvey, R. G. *J. Org. Chem.* **1979**, *44*, 4813-4816.
43. Reference 2, p.369.
44. Mallory, F. B.; Mallory, C. W. *Org. Reactions*, **1984**, *30*, 1-456.
45. Harvey, R. G.; Cortez, C.; Sugiyama, T.; Ito, Y.; Sawyer, T. W.; DiGiovanni, J. *J. Medic. Chem.* **1988**, *31*, 154-159.

Chapter 9

Chemistry of Polycyclic Aromatic Hydrocarbons (PAH) *o*-Quinones Generated by the Aldo-Keto Reductase Pathway of PAH Activation

Sridhar R. Gopishetty[1], Ronald G. Harvey[2], Seon-Hwa Lee[1], Ian A. Blair[1], and Trevor M. Penning[1]

[1]Department of Pharmacology, School of Medicine, University of Pennsylvania, Philadelphia, PA 19104
[2]The Ben May Institute for Cancer Research, University of Chicago, Chicago, IL 60637

Carcinogenic polycyclic aromatic hydrocarbons (PAHs) can be metabolically activated by human aldo-keto reductases (AKR) to three classes of *o*-quinones of different reactivity and redox-activity. Class I *o*-quinone [e.g. naphthalene-1,2-dione and 7,12-dimethylbenz[*a*]anthracene-3,4-dione (DMBA -3, 4-dione)] are among the most reactive AKR products and have the potential to form conjugates with cellular nucleophiles and form stable and depurinating adducts with DNA. We now show that DMBA-3,4-dione will react with 2-mercaptoethanol to yield, *mono-* and *bis-* thioether conjugates by undergoing sequential 1,6- and 1,4- Michael addition reactions. Similar reactions should occur with cellular GSH. We also describe methods for the synthesis of stable naphthalene-1,2-dione-N^2-dGuo and N^6-dAdo adducts which may be applicable to other PAH *o*-quinones.

Polycylic aromatic hydrocarbons (PAHs) are among the most ubiquitous carcinogens (*1*). Among the various PAHs contaminating the environment, methylated derivatives of the parent compound are the more potent carcinogens (*2*). The concentration of methylated PAHs in tobacco smoke and in emissions from certain fuel processes are often in the same range as those of some unsubstituted PAHs.

PAHs are procarcinogens and must be metabolically activated to electrophiles to exert their genotoxic effects. Three principal pathways have been proposed for PAH activation and are shown for the representative compound benzo[*a*]pyrene (Figure 1).

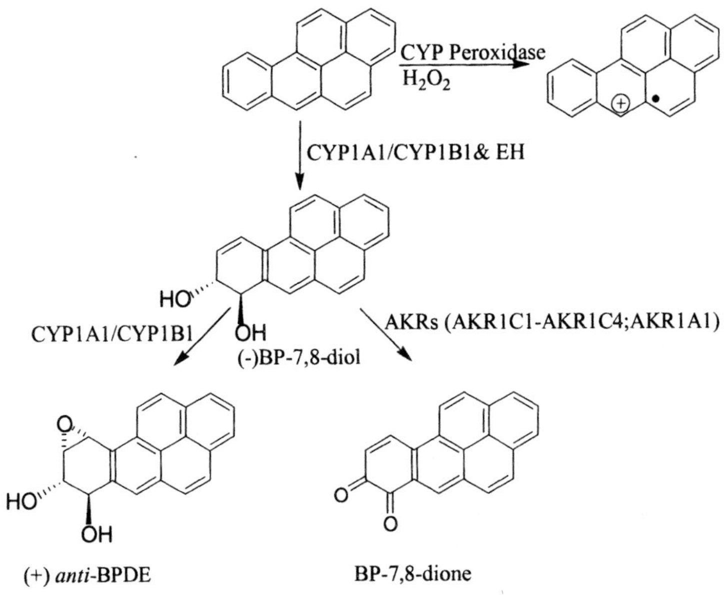

Figure 1. Three principal pathways for PAH activation; CYP = cytochrome P450 isoform; EH = epoxide hydrolase.

The first pathway involves the formation of radical cations catalyzed by P450 peroxidases (*3*). In the second pathway members of the CYP superfamily form an arene oxide on the terminal benzo-ring, subsequent hydrolysis by epoxide hydrolase results in the formation of non-K-region *trans*-dihydrodiols, which are potent proximate carcinogens (*4*). The *trans*-dihydrodiols can then undergo a secondary epoxidation to form bay region *anti* or *syn*-diol epoxides, which are potent mutagens and tumorigens (*5*). The third route involves dihydrodiol dehydrogenase(s) members of aldo-keto reductase (AKR) superfamily, which oxidize PAH *trans*-dihyrodiol to the corresponding *o*-quinone (*6-8*). The NADP$^+$ dependent oxidation of the *trans*-dihydrodiol initially results in a ketol, which spontaneously rearranges to form a catechol. The catechol is unstable and undergoes autoxidation in air to give rise to the fully oxidized *o*-quinone with the concomitant production of ROS (*9*) (Figure 2).

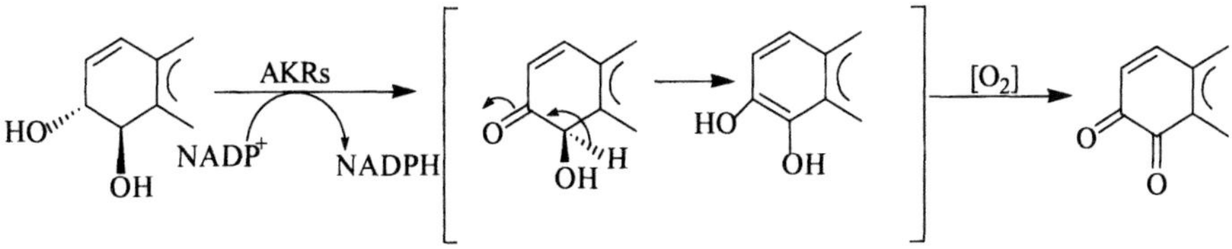

Figure 2. PAH activation by aldo-keto reductase(s).

The resulting *o*-quinones are highly reactive Michael acceptors, which can form conjugates with cellular nucleophiles and form stable and depurinating adducts with DNA (*10-11*) (Figure 3).

Earlier we reported the synthesis of the glutathionyl conjugate of benzo[*a*]pyrene-7,8-dione [*12*]. However, the chemistry of thiol addition to DMBA-3,4-dione, one of the most reactive *o*-quinones produced from the AKR pathway of PAH activation, has not been described. Also, synthesis of depurinating adducts of PAH *o*-quinones with deoxyguanosine has been reported [*11*]. But the synthetic standards required to validate the formation of stable PAH *o*-quinone DNA adducts in biological systems are still lacking. We now report the chemistry of these addition reactions.

Experimental Procedures

Caution

The work described involves the synthesis and handling of hazardous agents and was therefore conducted in accordance with the NIH Guidelines for the Laboratory Use of Chemical Carcinogens.

Materials

4-Bromo-naphthalene-1,2-dione and 4-amino-naphthalene-1, 2-dione were synthesized according to published procedures (*13-14*). Protected deoxyguanosine and deoxyadenosine were synthesized based on reported methods (*15*). Dimethylbenz[*a*]anthracene-3,4-dione was purchased from the

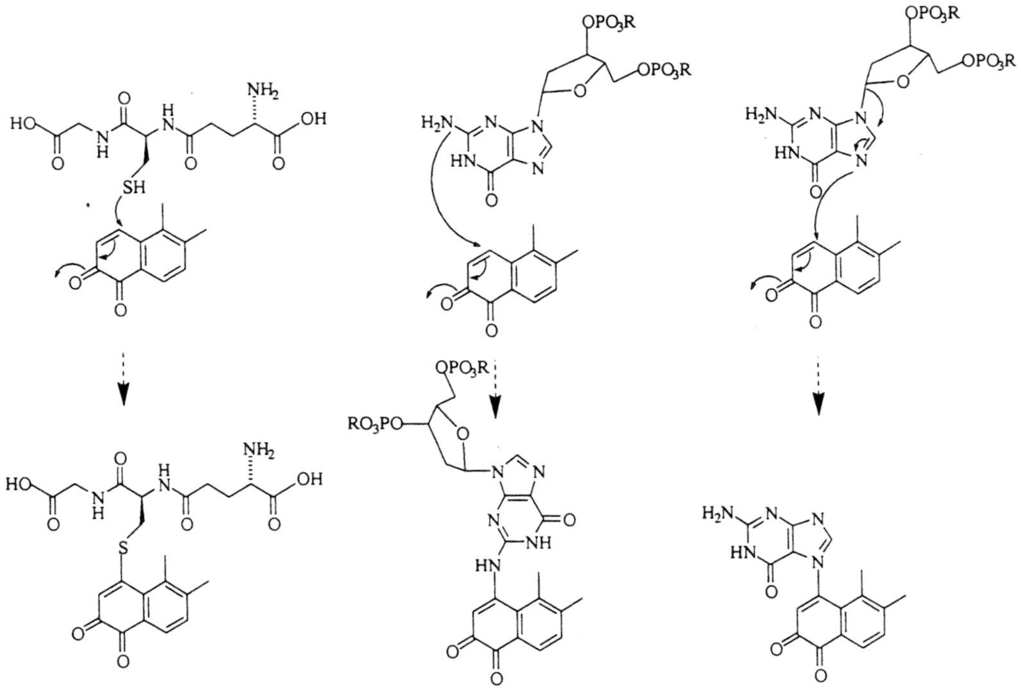

Figure 3. *Cellular conjugates and DNA adducts anticipated with PAH o-quinones.*

National Cancer Institute Chemical Carcinogen Reference Standard Repository at Chemsyn Science LABS (Lenexa, KS). All other chemicals and solvents were purchased from Sigma Chemical Co. (St. Louis, MO) and used without further purification.

HPLC

Mercaptoethanol conjugates were purified using a semi-preparative RP-HPLC column (Partisil 10 ODS, 9.5 x 500 mm; Whatman, Clifton, NJ) developed isocratically with 50:50 acetonitrile:water mobile phase at flow rate of 3 mL/min using a Beckman System Gold HPLC.

Mass Spectrometry

Mass spectrometric data for mercaptoethanol conjugates were acquired on a Finnigan LCQ ion trap mass spectrometer (ThermoQuest, San Jose, CA) equipped with a Finnigan electrospray ionization (ESI) source. The mass spectrometer was operated in the positive-ion mode. On-line chromatography was performed using a Waters Alliance 2690 HPLC system (Waters Corp, Milford, MA). A YMC C_{18} ODS-AQ column was used at a flow rate of 0.9 mL/min. Solvent A was 5 mM ammonium acetate in water containing 0.01% trifluoroacetic acid and solvent B was 5 mM ammonium acetate in methanol containing 0.01% trifluoroacetic acid with the gradient conditions as follows: 30% B at 0 min, 30% B at 5 min, 100% B at 16 min, 100% B at 24 min and 30% B at 26 min. Similarly mass spectrometric data for stable DNA adducts were acquired using an Alltech CN column, at flow rate of 1.0 mL/min. Solvent A was Hexane/Isopropanol (197:3) and solvent B was Hexane/Isopropanol (70:30) with the gradient conditions as follows: 1% B at 0 min, 1% B at 5 min, 25% B at 10 min, 25% B at 15 min and 1% B at 20 min.

NMR Spectroscopy

^{1}H NMR spectra were obtained on a Varian Unity Spectrometer operating at 500 MHz, using $CDCl_3$ as the solvent. Chemical shifts are relative to TMS.

Synthesis of Thio-ether Conjugates of DMBA-3,4-dione

A 50 mL reaction containing potassium phosphate buffer (10 mM, pH 7.0), 5 mM 2-mercaptoethanol and 25 μM DMBA-3,4-dione (1) in 8% DMSO was stirred for 20 h at 37 °C (Figure 4).

The reaction mixture was extracted with ethyl acetate. Organic extracts were dried and concentrated under reduced pressure to yield a brown solid, which was purified by semi preparative HPLC using water/acetonitrile. The products obtained after purification were characterized by LC/MS and NMR and assigned to the *mono-* and *bis* thio-ether conjugates, **2**, **3**.

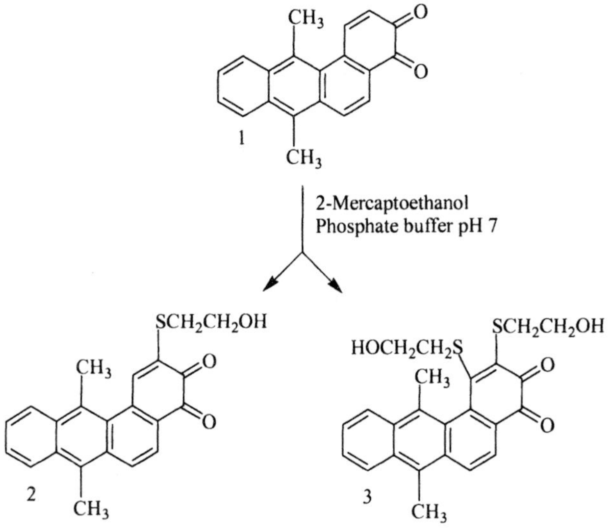

Figure 4. Synthesis of mono- and bis-thioether conjugates of DMBA-3, 4-dione.

Synthesis of *bis*-TBDMS-O^6-NPE naphthalene-1,2-dione-N^2-dGuo (6) and *bis*-TBDMS-naphthalene-1,2-dione-N^6-dAdo (8)

An oven dried reaction vial was charged with *bis*-TBDMS-O^6-nitrophenethyl-dGuo (6.4 mg) **(5)**, cesium carbonate (4.5 mg), palladium acetate (2.3 mg), BINAP (9 mg), 4-bromo-1,2-naphthoquinone **(4)** in toluene (2 mL) (Figure 5). The vial was flushed with argon. The reaction mixture was stirred for 30 min at room temperature, heated at 80 °C for 16 h, and then diluted with ethyl acetate (10 mL) and clarified by centrifugation (*16*). The supernatant liquid afforded **(6)**, which was dissolved in acetonitrile and further analysis was carried out by LC/MS. Similarly *bis*-TBDMS-dAdo **(7)** was coupled with 4-bromo-1,2-naphthoquinone to give **(8)**.

Synthesis of O^6-NPE naphthalene-1, 2-dione-N^2-dGuo (11) and naphthalene-1, 2-dione-N^6-dAdo (13)

Deoxy-2-fluoro-O^6-NPE-inosine (10) (5 mg) and 4-amino-naphthalene-1, 2-dione (9) were dissolved in anhydrous DMSO (0.5 mL) followed by addition of diisopropylethyl amine (25 μL) (Figure 6). The reaction was stirred for 5 h at 45 °C (17), and then concentrated *in vacuo* to yield (11).

Coupling of 6-chloropurine (12) with 4-amino-naphthalene-1,2-dione was conducted in a similar manner to yield (13).

Results and Discussion

Formation of DMBA-3, 4-dione Conjugate with 2-Mercaptoethanol.

o-Quinones are highly reactive Michael acceptors that have the propensity to form conjugates via 1,4-Michael addition. Of the quinones anticipated to form by AKRs, DMBA-3, 4-dione is among the most reactive. DMBA-3, 4-dione was reacted with 5 mM 2-mercaptoethanol in phosphate buffer (pH 7.0) (Figure 4). Analysis by reversed-phase HPLC indicated that virtually all of the DMBA-3, 4-dione was consumed and led to the formation of two products more polar than DMBA-3, 4-dione with retention times of 27.7 min and 60.5 min respectively.

Analysis by HPLC/electrospray ionization/mass spectrometry (HPLC/ESI/ MS) in conventional full scan mode showed a protonated molecular ion (MH$^+$) at m/z 439, for peak at R$_t$=27.7. Based on the MH$^+$ and fragment ions m/z=421 [MH-H$_2$O], m/z= 362 [MH-SCH$_2$CH$_2$OH] this peak is characterized as a *bis*-thioether conjugate of DMBA-3, 4-dione (3). Similarly the peak at R$_t$=60.5 gave molecular ion (MH$^+$) at m/z 363 and a fragment ion m/z= 345 [MH-H$_2$O] corresponding to the *mono*-thioether conjugate of DMBA-3, 4-dione (2). ^{1}H NMR of the *mono*-thioether conjugate showed the absence of the vinylic proton at C2 (6.37 ppm) demonstrating that this is the initial site of thiol attack.

In previous studies we have shown that naphthalene-1,2-dione and benzo [*a*]pyrene-7.8-dione reacted with 2-mercaptoehtanol to give only one product and this corresponded to the 1,4-Micheal addition products of the fully oxidized *o*-quinone [8]. By contrast, the products obtained from addition of 2-mercaptoetanol to DMBA-3, 4-dione, correspond to a *mono*-thioether conjugate that arises from 1,6-Michael addition and a *bis*-thioether conjugate formed by 1,4-Michael addition to the *mono*-thioether conjugate. It is proposed that the first Michael addition occurs via 1,6-addition through the ring system and this is supported by the NMR data (Figure 7).

The ability to form the *bis*-thioether conjugate may explain why previous attempts to chemically synthesize a simple thioether conjugate of DMBA-3, 4-dione failed. Importantly, the detection of the *bis*-conjugate raises the issue of

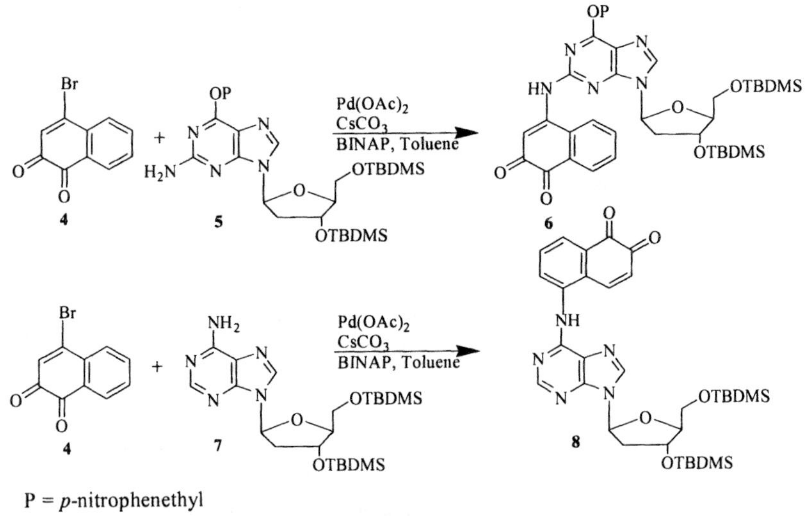

P = *p*-nitrophenethyl

Figure 5. Synthesis of naphthalene-1,2-dione stable adducts via bromo-o-quinones: Method I

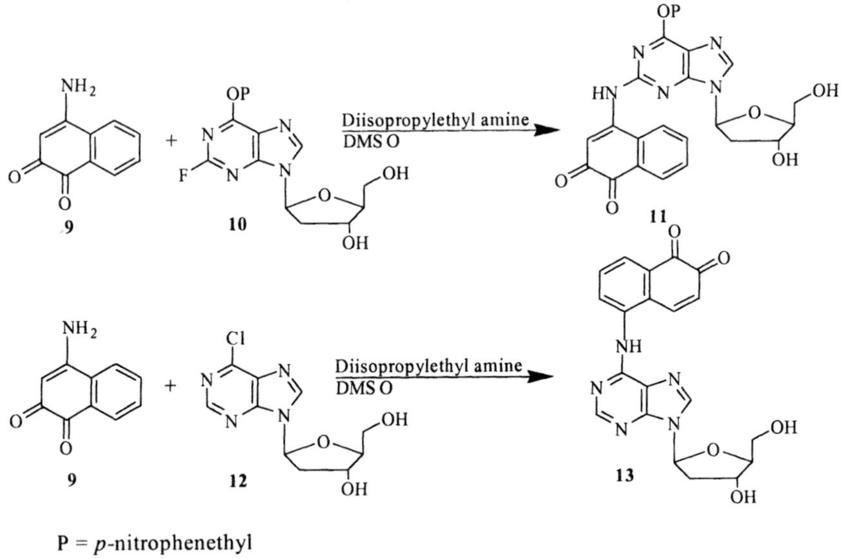

P = *p*-nitrophenethyl

Figure 6. Synthesis of naphthalene-1, 2-dione stable adducts via amino-o-quinones: Method II

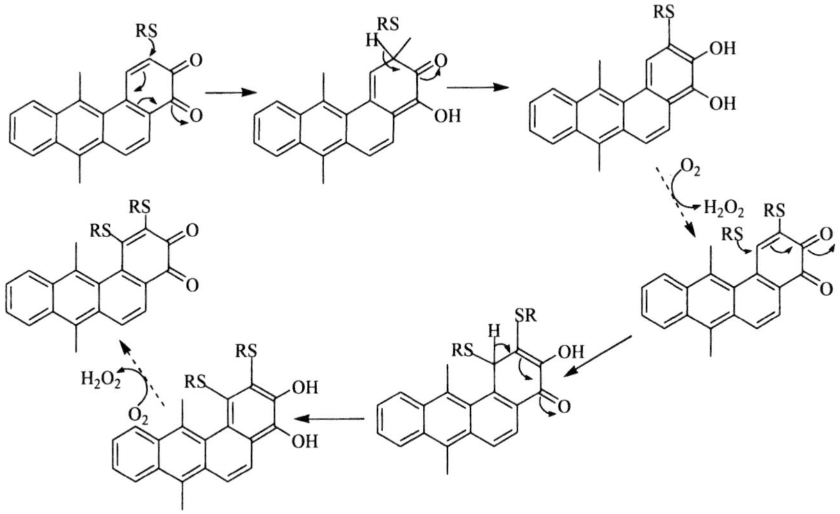

Figure 7. Mechanism of mono- and bis-thioether conjugate formation of DMBA-3, 4-dione.

whether other cellular nucleophiles could form similar *bis*-conjugates e.g. GSH. *bis*-Glutathionyl ether conjugates of variety of polyphenols have been implicated in hematotoxicity and nephrotoxicity, raising the prospect that *bis*-thioether conjugates of DMBA-3,4-dione may have their own toxicological profile (*18*).

Synthesis of Naphthalene-1,2-dione Stable Adducts with Protected dGuo and dAdo (6) and (8) via Bromo-*o*-quinones: Method I

Two options were considered for the synthesis of naphthalene-1,2-dione stable adducts. The first method involves coupling bromo-*o*-quinones with protected deoxyribonucleosides (Figure 5) using the conditions of the Buchwald-Hartwig reaction. This method is similar to the procedure that Hopkins (*19*) and Johnson (*20*) used for the synthesis of the cross-linked adducts of dG. The protected deoxyguanosine derivative (5) was prepared as previously described (*15*). The coupling of bromo-*o*-quinone (4) and deoxyguanosine (5) was achieved in the presence of palladium acetate as catalyst and the reaction was carried out at 80 °C in toluene. The product was characterized by HPLC/electrospray ionization/mass spectrometry (HPLC/ESI/MS). LC/MS analysis showed the fragment ions at m/z 687 [MH-TBDMS]+H, m/z=644, m/z=494 and m/z=536, based on mass spectral data the compound was characterized as (6).

Similarly the stable adduct of protected deoxyadenosine was synthesized and LC/MS analysis showed the fragment ions m/z=522 [MH-TBDMS]+H, m/z=288 [M-*bis*-TBDMS-dR], m/z=178 derived from (8).

Synthesis of Naphthalene-1,2-dione Stable Adducts with Halogenated dGuo and dAdo (11) and (13) via Amino-*o*-quinones: Method II

The second method involved coupling halogenated deoxyribonucleosides with amino *o*-quinones. The fluoro deoxyribonucleoside (10) was prepared as previously described (*21*) and the amino-*o*-quinone was dissolved in DMSO. The reaction was initiated by the addition of diisopropylethyl amine and carried out at 45 °C for 5 h. The product was characterized by HPLC/electrospray ionization/mass spectrometry (HPLC/ESI/MS). LC/MS analysis showed the fragment ions at m/z 572 [MH$^+$] and m/z=555 [MH-H$_2$O]+H, based on the mass spectral data the compound was identified as (11). Similar reactions with 6-chloropurine and amino-*o*-quinone failed to yield product.

One goal of this work was to synthesize PAH-*o*-quinone stable adducts as chemical standards to detect the prevalence of these adduct in biological systems. Stable adducts of protected dGuo and dAdo with naphthalene-1,2-dione were prepared and characterized by LC/MS. The structural similarity between naphthalene-1,2-dione and other PAH *o*-quinones and estrogen *o*-quinones suggests that the synthetic routes described may be applicable to form stable adducts of these more biologically relevant *o*-quinones. Identification of these adducts in biological systems is part of an ongoing study designed to evaluate the contribution of PAH *o*-quinones in the metabolic activation of PAHs. Importantly, stable adducts formed by reaction of PAH *o*-quinones with DNA could provide a route to the mutations observed in human cancers that arise from exposure to PAHs.

Acknowledgment

This work was supported by RO1 grants CA39504 and PO1 CA92537 (to T.M.P) from the National Cancer Institute.

References

1. *Polycyclic Hydrocarbons and Carcinogenesis;* Dipple, A.; Harvey, R.G., Eds.; American Chemical Society: Washington, DC, 1985; p 1-18.
2. *Chemical Carcinogens;* Dipple, A.; Searle, C. E., Eds.; American Chemical Society: Washington, DC, 1976; p 245-314.

3. Cavalieri, E. L.; Rogan, E. G. *Xenobiotica,* **1995**, *25*, 677-688.
4. Morrison, V. M.; Burnett, A. K.; Forrester, L. M.; Wolf, C. R.; Craft, J. A. *Chem-Biol. Interactions*, **1991**, *79*, 179-196.
5. Gelbion, H. V.; *Physiological Reviews*, **1980**, *60*, 1107-1166.
6. Smithgall, T. E.; Harvey, R. G.; Penning, T. M. *J. Biol. Chem.* **1986**, *261*, 6184-6191.
7. Smithgall, T. E.; Harvey, R. G.; Penning, T. M. *Cancer Res.* **1988**, *48*, 1227-1232.
8. Smithgall, T. E.; Harvey, R. G.; Penning, T. M. *J. Biol. Chem.* **1988**, *263*, 1814-1820.
9. Penning, T. M.; Ohnishi, S. T.; Ohnishi, T.; Harvey, R. G. *Chem. Res. Toxicol,* **1996**, *9*, 84-92.
10. Shou, M.; Harvey, R. G.; Penning, T. M. *Carcinogenesis*, **1993**, *14*, 475-482.
11. McCoull, K. D.; Rindgen, D.; Blair, I. A.; Penning, T. M. *Chem. Res. Toxicol.* **1999**, *12*, 237-246.
12. Murthy, V. S.; Penning, T. M. *Bioconjugate Chemistry.* **1992**, *3*, 218-224.
13. Harinath, B.; Subbarao, N. V. *Proc. Indian. Acad. Sci. Sec A.* **1967**, *66*, 301-305.
14. Husu, B. K.; Kadunc, Z.; Tisler, M. *Monatshefte fur Chemie.* **1988**, *119*, 215-221.
15. Jinsuk, W.; Snorri, T. S.; Hopkins, P. B. *J. Am. Chem. Soc.* **1993**, *115*, 3407-3415.
16. Riccardis, F. D.; Bonala, R. R.; Johnson, F. *J. Am. Chem. Soc.* **1999**, *121*, 10453-10460.
17. Hye-Young, H. K.; Monica, C.; Lubomir, V. N.; Constance, M.H.; Thomas, M. H. *Chem. Res. Toxicol.* **2001**, *14*, 1306-1314.
18. Monks, T. J.; Lau, S.S. *Chem. Res. Toxicol.* **1997**, *10*, 1296-1313.
19. Harwood, E. A.; Sigurdsson, S. T.; Edfeldt, N. B. F.; Reid, B. R.; Hopkins, P. B. *J. Amer. Chem. Soc.* **1999**, *121*, 5081-5082
20. Bonala, R. R.; Shishkina, I. G.; Johnson, F. *Tet. Letters*, **2000**, *41*, 7281-7287.
21. Acedo, M.; Fabrega, C.; Avino, A.; Goodman, M.; Fagan, P.; Wemmer, D.; Eritja, R. *Nucleic Acids Research*, **1994**, *22*, 2982-2989.

Chapter 10

Analysis of Etheno 2'-Deoxyguanosine Adducts as Dosimeters of Aldo-Keto Reductase Mediated-Oxidative Stress

Seon Hwa Lee and Ian A. Blair

Center for Cancer Pharmacology, University of Pennsylvania, Philadelphia, PA 19104

The formation of polyunsaturated fatty acid lipid hydroperoxides is mediated by reactive oxygen species such as those formed as a consequence of AKR-mediated polycyclic aromatic hydrocarbon catechol formation. Lipid hydroperoxides undergo homolytic decomposition to the bifunctional electrophiles, 4-hydroperoxy-2-nonenal, 4-oxo-2-nonenal, 4-hydroxy-2-nonenal, and 4,5-epoxy-2-decenal. The major DNA-adducts resulting from reaction with these lipid hydroperoxide decomposition products are etheno-2'-deoxyguanosine and heptan-2-one-etheno-2'-deoxyguanosine. Specific derivatization of these adducts with pentafluorobenzyl bromide facilitated their analysis by liquid chromatography/atmospheric pressure chemical ionization/tandem mass spectrometry. An extraction method and purification procedure was developed so that these adducts could be analyzed as pentafluorobenzyl derivatives in human urine using stable isotope dilution methodology.

Polyunsaturated fatty acid (PUFA) lipid hydroperoxides are formed non-enzymatically by reactive oxygen species (ROS) such as superoxide ($O_2^{-\bullet}$), peroxide (O_2^{2-}), and hydroxyl radical ($HO^\bullet$). The endogenous pathways for ROS generation include normal mitochondrial aerobic respiration, phagocytosis of bacteria or virus-containing cells, peroxisomal-mediated degradation of fatty acids, and cytochrome P450-mediated metabolism of xenobiotics (*1*). AKR-mediated oxidation of PAH dihydrodiols to catechols and subsequent conversion to the corresponding quinones can also result in the formation of ROS (*2*) (Figure 1). Antioxidant defense systems *in vivo* that can detoxify ROS include:

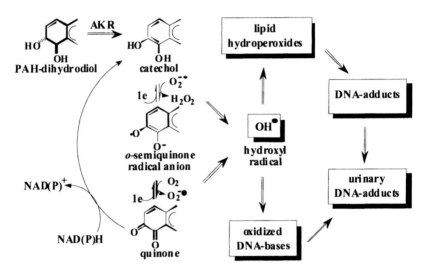

Figure 1. AKR-mediated formation of Reactive Oxygen Species (ROS), lipid hydroperoxides, and DNA-adducts

superoxide dismutase, catalase, and reduced glutathione (GSH)-dependent peroxidases (*1*). Also, endogenous processes such as the sequestration of hydrogen peroxide generating enzymes or the chelation of free transition metal ions by transferrin, ferritin, and ceruloplasmin can protect against ROS-mediated damage. However, it is always possible that cellular macromolecules and lipids can be damaged by the ROS that escapes from these defense systems. For example, PAH can initiate damage to DNA through a redox cycle involving ROS, which is implicated in lung carcinogenesis (*3*). ROS-mediated formation of lipid hydroperoxides is a complex process, which involves initiation by the abstraction of a *bis*-allylic methylene hydrogen atom of the PUFA followed by addition of molecular oxygen (*4*). This results in the formation of 9- and 13-hydroperoxy-octadecadienoic acid (HPODE) isomers from linoleic acid, the major ω-6 PUFA present in plasma lipids.

PUFA lipid hydroperoxides can also be formed enzymatically from lipoxygenases (LOXs) (5) and cyclooxygenases (COXs) (6) with much greater stereoselectivity than is observed in a free radical mechanism. Up-regulation of COX-2 has been observed in colon cancer (7); whereas up-regulation of 12-LOX has been observed in skin models of PAH carcinogenesis (8). Human 12-LOX convert linoleic acid to mainly 13-[S-(Z,E)]-9,11-HPODE [13(S)-HPODE] and 9-[R-(E,Z)]-10,12-HPODE [9(R)-HPODE] (9). COX-1 and COX-2 also produce 13(S)-HPODE and 9(R)-HPODE (10). The other C-18 PUFAs including linolenic acid (ω-3) and dihomo-γ-linolenic acid (ω-6) and all C-20 PUFAs also undergo LOX-mediated conversion to hydroperoxides. Products that arise from COX-mediated metabolism of C-20 PUFAs are prostaglandins, and thromboxanes, rather than lipid hydroperoxides.

PUFA lipid hydroperoxides undergo transition metal ion- or vitamin C-induced decomposition to α,β-unsaturated aldehyde genotoxins that can react with DNA (11). We have identified 4-oxo-2-nonenal as one of the major products from the homolytic decomposition of 13-HPODE (a prototypic ω-6 PUFA lipid hydroperoxide) (12) (Figure 2).

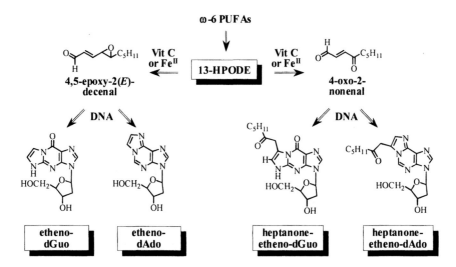

Figure 2. Formation of etheno adducts

Spiteller *et al.* also confirmed the formation of 4-oxo-2-nonenal from 13-HPODE (13). 4-Hydroperoxy-2-nonenal was also characterized as a product of 13-HPODE decomposition (12,14). Liquid chromatography (LC)/atmospheric pressure chemical ionization (APCI)/mass spectrometry (MS) methodology was used to identify the α,β-unsaturated aldehydic bifunctional electrophiles that were formed during homolytic lipid hydroperoxide decomposition (*15*). Using this methodology, *trans*-4,5-epoxy-2(*E*)-decenal (major) and *cis*-4,5-epoxy-2(*E*)-decenal (minor) were also identified. It has also been demonstrated that 4-hydroperoxy-2-nonenal is a precursor in the formation of 4-oxo-2-nonenal and 4-hydroxy-2-nonenal (*5*).

The reaction between 4-oxo-2-nonenal and 2'-deoxyguanosine (dGuo) results in the initial formation of three unstable adducts that subsequently dehydrate to form a single adduct (*16*). The three initially formed ethano adducts arise from highly regioselective nucleophilic addition of N^2 of the dGuo to the C1 aldehyde of 4-oxo-2-nonenal followed by reaction of N1 at C2 of the resulting α,β-unsaturated ketone. All three ethano adducts then dehydrate to a single heptan-2-one-etheno-dGuo adduct (H-dGuo; Figure 2). In contrast, two DNA-adducts are formed initially in the reaction between 2'-deoxyadenosine (dAdo) and 4-oxo-2-nonenal (*17*). The initially formed adducts then each decompose to give a single adduct. These data are consistent with nucleophilic addition of N^6 of dAdo to the C1 aldehyde of 4-oxo-2-nonenal followed by reaction of N1 at C2 of the resulting α,β-unsaturated ketone to generate a mixture of two ethano adducts that inter-convert with each other. The ethano adducts subsequently dehydrate to give a single heptan-2-one-etheno-dAdo adduct (H-dAdo'; Figure 2).

Trans,trans-2,4-Decadienal is an α,β-unsaturated aldehydic decomposition product from 9-hydroperoxy-(*E,E*)-10,12-octadecadienoic acid or 11-hydroperoxy-(*Z,Z,E,E*)-eicosa-5,8,12,14-tetraenoic acid (*13,18*). Recent studies have shown that the reaction of peroxide-treated *trans,trans*-2,4-decadienal with dAdo or dGuo results in the formation of 1,N^6-etheno-dAdo (E-dAdo) (*19*) and 1,N^2-etheno-dGuo (E-dGuo) (*20*), respectively. We reasoned that *trans*-4,5-epoxy-2(*E*)-decenal could have been formed when *trans,trans*-2,4-decadienal was treated with peroxides (*18,21*) and subsequently showed that this bifunctional electrophile is in fact the precursor to the formation of E-dGuo from the homolytic decomposition of PUFA lipid hydroperoxides (*22*) (Figure 2).

We have now developed methodology for the analysis of E-dGuo and H-dGuo based on stable isotope dilution LC/APCI/tandem mass spectrometry (MS/MS).

Materials and Methods

Materials

Ammonium acetate, ammonium hydroxide, ascorbic acid, chloroacetaldehyde, dGuo, diisopropylethylamine amine (DIPE), formic acid, 2,3,4,5,6-pentafluorobenzyl bromide (PFB-Br) were purchased from Sigma-Aldrich. (St. Louis, MO). 3-Morpholinopropanesulfonic acid (MOPS) was obtained from Fluka BioChemika (Milwaukee, WI). $^{15}N_5$-dGuo was obtained from Cambridge Isotope Laboratories (Andover, MI). Supelclean LC-18 and LC-Si solid-phase extraction (SPE) columns were from Supelco (Bellefonte, PA). Chelex-100 chelating ion exchange resin (100-200 mesh size) was purchased from Bio-Rad Laboratories (Hercules, CA). HPLC grade water, acetonitrile, methanol, methylene chloride were obtained from Fisher Scientific Co. (Fair Lawn, NJ). Gases were supplied by BOC Gases (Lebanon, NJ).

Liquid Chromatography

Chromatography for LC/MS experiments was performed using a Waters Alliance 2690 HPLC system (Waters Corp., Milford, MA). Purification of $^{15}N_5$-analogs for dGuo-adducts was conducted on a Hitachi L-6200 Intelligent Pump (Hitachi, San Jose, CA) equipped with a Hitachi L-4000 UV detector. Gradient elution was performed in the linear mode. A Synergi Polar-RP column (250 x 4.6 mm i.d., 4 μm; Phenomenex, Torrance, CA) was employed for gradient system 1 with a flow rate of 1.0 mL/min. Systems 2 employed an YMC basic column (150 x 2.0 mm i.d., 5 μm; Waters, Milford, MA), system 3 employed YMC-Pack ODS-AQ column (250 x 10 mm i.d., 5 μm) and system 4 and 5 employed YMC-Pack ODS-AQ column (250 x 4.6 mm i.d., 5 μm). For system 1, solvent A was 5 mM ammonium acetate in water and solvent B was 5 mM ammonium acetate in acetonitrile. The linear gradient for system 1 was as follows: 6% B at 0 min, 6% B at 3 min, 20% B at 9 min, 20% B at 13 min, 60% B at 22 min, 80% B at 27 min, 80% B at 32 min. The separation was performed at 30 °C. For systems 2, solvent A was 5 mM aqueous ammonium acetate containing 0.01 % (v/v) trifluoroacetic acid and solvent B was 5 mM ammonium acetate containing 0.01% (v/v) trifluoroacetic acid in methanol. The linear gradient for system 2 was as follows: 1% B at 0 min, 91% B at 18 min, 91% B at 20 min with a flow rate of 0.25 mL/min. System 3 employed an isocratic mobile phase of methanol/water (1:1, v/v; 2 mL/min). For systems 4, solvent A was water containing 0.01% (v/v) trifluoroacetic acid and solvent B was acetonitrile containing 0.01% (v/v) trifluoroacetic acid. For systems 5, solvent A was water and solvent B was acetonitrile. The linear gradient for system 4 and 5 was as follows: 20% B at 0 min, 20% B at 3 min, 35% B at 9 min, 35% B at 13 min, 60% B at 21 min with a flow rate of 1 mL/min. All separations except for system 1 were performed at ambient temperature.

Mass Spectrometry

Mass spectrometry was conducted with a Thermo Finnigan LCQ ion trap mass spectrometer or a Thermo Finnigan TSQ 7000 triple-stage quadrupole mass spectrometer (Thermo Finnigan, San Jose, CA) equipped with an APCI source in positive ion mode. The LCQ operating conditions were as follows: vaporizer temperature at 450 °C, heated capillary temperature at 150 °C, with a discharge current of 5 µA applied to the corona needle. Nitrogen was used as the sheath (80 psi) and auxiliary (10 arbitrary units) gas to assist with nebulization. The TSQ 7000 operating conditions were as follows: vaporizer temperature at 550 °C, heated capillary temperature at 160 °C, with the corona discharge needle set at 8 µA. The sheath gas (nitrogen) and auxiliary gas (nitrogen) pressures were 80 psi and 3 (arbitrary units), respectively.

Vitamin C-Mediated Decomposition of 13-HPODE in the Presence of dGuo

A solution of 13-HPODE (150 nmol) in ethanol (10 µL) and vitamin C (750 nmol) in Chelex-treated 100 mM MOPS containing 150 mM NaCl (pH 7.4, 10 µL) were added to dGuo (750 nmol) in Chelex-treated 100 mM MOPS containing 150 mM NaCl (pH 7.4, 180 µL). The reaction mixture was sonicated for 15 min at room temperature, incubated at 37 °C for 24 h and then placed on ice. The sample was filtered through a 0.2 µm Costar cartridge prior to analysis of a portion of the sample (20 µL) by LC/MS using gradient system 1.

Derivatization

E-dGuo and H-dGuo in acetonitrile (100 µL) were treated with 20% PFB-Br in acetonitrile (v/v, 100 µL) and 10% DIPE in acetonitrile (v/v, 100 µL) at room temperature for 2 h. The reaction mixture was evaporated to dryness under nitrogen and re-dissolved in acetonitrile/water (8:2, v/v) ready for LC/MS analysis.

LC/MS Analysis of PFB-Derivatives

LC/APCI/MS analysis for the PFB-derivatives of dGuo-adducts was conducted using LCQ ion trap mass spectrometer with gradient system 1. The full scan analyses were performed in the range of m/z 50 to m/z 800.

Preparation of $^{15}N_5$-E-dGuo and $^{15}N_5$-H-dGuo

$^{15}N_5$-E-dGuo was prepared by a modification of the method of Sattsangi et al. (23) from the reaction of $^{15}N_5$-dGuo with chloroacetaldehyde at 37 °C at pH

6.4 for 5 days in water. After purification by reversed phase HPLC using system 2 (t_R = 12 min) followed by system 3 (t_R = 8 min), $^{15}N_5$-E-dGuo was obtained as a white solid (U λ_{max} = 226, 285 nm in methanol/water containing 0.1% trifluoroacetic acid) with > 97% purity by HPLC system 2 (t_R = 12 min). APCI/MS: m/z 297 (MH$^+$), m/z 181 (BH$_2^+$). MS^2 (m/z 297→): m/z 181 [BH$_2$]$^+$. $^{15}N_5$-H-dGuo was prepared by the reaction of $^{15}N_5$-dGuo with 4-oxo-2-nonenal at 60 °C for 24 h in water. It was placed on ice for 15 min and $^{15}N_5$-H-dGuo was purified using gradient system 4 (t_R = 9.6 min), followed by system 5 (t_R = 11.10 min) by monitoring the UV absorbance at 231nm. $^{15}N_5$-H-dGuo was obtained as a white solid (UV λ_{max}: 229 nm at pH 7, 236 nm at pH 13, and 226 at pH 1) with > 98% purity by HPLC system 5 (t_R = 11.1 min). APCI/MS: m/z 409 (MH$^+$), m/z 293 (BH$_2^+$). MS^2 (m/z 409→): m/z 293 [BH$_2$]$^+$.

MS/MS Analysis of ^{14}N- and ^{15}N-dGuo-Adducts PFB-Derivatives

MS/MS analysis was performed on PFB derivatives of ^{14}N- and ^{15}N-dGuo by LCQ ion trap mass spectrometer. Helium was employed as a collision gas and the relative collision energy was set at 1 V (20% of maximum).

LC/MRM/MS analysis of PFB-Derivatives

LC/multiple reaction monitoring (MRM)/MS analysis was conducted using 10 ng of ^{14}N- and ^{15}N-dGuo-adducts as its PFB derivatives. TSQ 7000 triple-stage quadrupole mass spectrometer was used for this experiment with gradient system 1. Collision-induced dissociation (CID) was performed using argon as the collision gas at 2.7 mTorr in the second (rf-only) quadrupole. For the MRM analysis, unit resolution was maintained for both parent and product ions. The following MRM transitions were monitored: E-dGuo-PFB; m/z 472 → m/z 356 (collision energy: –18 eV), $^{15}N_5$-E-dGuo-PFB; m/z 477 → m/z 361 (collision energy: –18 eV), H-dGuo-PFB; m/z 584 → m/z 468 (collision energy: –20 eV), and $^{15}N_5$-H-dGuo-PFB; m/z 589 → m/z 473 (collision energy: –20 eV).

Extraction from Urine

$^{15}N_5$-E-dGuo and $^{15}N_5$-H-dGuo (100 ng of each) were added to urine sample (6 mL) and the solution was adjusted to pH 3 with 20% formic acid, Figure 3. DNA-adducts were extracted with 6 mL of ethyl acetate. The organic layer was then neutralized by adding 2 M NH$_4$OH and evaporated to dryness under nitrogen.

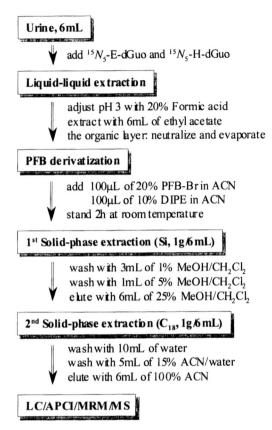

Figure 3. Overall procedure for analysis of etheno-DNA-adducts in urine

LC/MS Analysis of Urine Extracts

The tube containing $^{15}N_5$-E-dGuo, $^{15}N_5$-H-dGuo and DNA adducts in acetonitrile (100 µL) was treated with 20% PFB-Br in acetonitrile (v/v, 100 µL) and 10% DIPE in acetonitrile (v/v, 100 µL) at room temperature for 2 h. After the reaction mixture was evaporated to dryness under nitrogen, it was dissolved in 25% methanol in dichloromethane (v/v). The PFB derivatized dGuo-adducts were purified by a Si SPE column (1g, 6 mL, Supelco) preconditioned with 15 mL of dichloromethane. The column was washed with 3 mL of 1% methanol in dichloromethane (v/v) and 3 mL of 5% methanol in dichloromethane (v/v), and PFB derivatized dGuo-adducts were eluted with 6 mL of 25% methanol in dichloromethane (v/v). The eluent was evaporated and re-dissolved in 80%

acetonitrile in water (v/v). The solution was then loaded on a C18 SPE column (1g, 6 mL, Supelco) preconditioned with 15 mL of acetonitrile, followed by 15 mL of water. The cartridge was washed with water (10 mL) and 15% acetonitrile in water mixture (5 mL, v/v). dGuo-Adducts were eluted in acetonitrile (6 mL). The eluent was evaporated to dryness under nitrogen and re-dissolved in 80% acetonitrile in water (100 μL). LC/APCI/MRM/MS analysis was conducted on a 20 μL aliquot of this solution using gradient system 1 (Figure 3).

Results and Discussion

Vitamin C-Mediated Decomposition of 13-HPODE in the presence of dGuo

LC/MS analysis of the reaction mixture revealed two major adducts (Figure 4). The mass spectrum of the most polar adduct (10.4 min) revealed an intense MH$^+$ at m/z 292, together with a BH$_2^+$ ion at m/z 176. MS^2 analysis of MH$^+$ resulted in a BH$_2^+$ product ion. These LC/MS characteristics were identical to those for 1,N^2-etheno-dGuo (E-dGuo). LC/MS analysis of the most abundant adduct (20.6 min) showed an MH$^+$ at m/z 404, together with BH$_2^+$ at m/z 288. This corresponded to heptanone etheno-dGuo (H-dGuo).

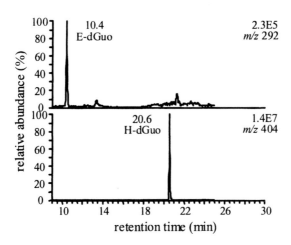

Figure 4. LC/MS analysis of E-dGuo and H-dGuo

LC/MS Analysis of PFB-Derivatives

PFB derivatives of E-dGuo and H-dGuo were formed in essentially quantitative yield. After PFB-derivatization, LC/MS analysis in the positive APCI mode showed the expected MH^+ ion at m/z 472 for E-dGuo and at m/z 584 for H-dGuo. PFB-derivatives of E-dGuo and H-dGuo were eluted with a higher proportion of organic solvent, which provided a better separation and higher sensitivity in comparison with those obtained from the un-derivatized adducts (Figure 5); retention times were changed from 10.4 min to 21.2 min (E-dGuo) and from 20.6 min to 24.7 min (H-dGuo) and the MS sensitivity of two adducts were increased more than ten-fold.

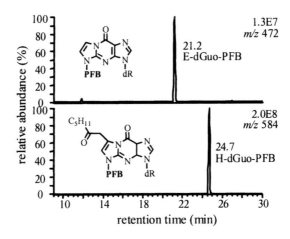

Figure 5. LC/MS analysis of E-dGuo-PFB, and H-dGuo-PFB

MS/MS Analysis of ^{14}N- and ^{15}N-dGuo Adducts PFB-Derivatives

MS^2 analysis of m/z 472 (MH^+ of E-dGuo-PFB) resulted in the exclusive formation of the BH_2^+ product ion at m/z 356, which is corresponds to the loss of 2′-deoxyribose with the addition of an hydrogen atom. The MS and MS^2 analysis of $^{15}N_5$-E-dGuo-PFB showed identical patterns of spectra to those for E-dGuo with MH^+ (m/z 477) and BH_2^+ (m/z 361) exhibiting a 5 Da increase in mass. MS^2 analysis of MH^+ of H-dGuo-PFB (m/z 584) gave rise to exclusive formation of the BH_2^+ product ion at m/z 468. Similarly, MS spectrum of $^{15}N_5$-H-dGuo-PFB showed an intense MH^+ at m/z 589 together with BH_2^+ at m/z 473 and MS^2 spectrum (m/z 589 →) exhibited BH_2^+ at m/z 473 as a major production.

LC/MRM/MS analysis of PFB-Derivatives

Analysis of E-dGuo-PFB and H-dGuo-PFB by LC/MRM/MS using gradient system 1 revealed that complete separation was accomplished between two dGuo-adducts (Figure 6). $^{15}N_5$-E-dGuo and $^{15}N_5$-H-dGuo were employed as internal standards. MRM analyses were performed using the transition from [MH$^+$] to [BH$_2^+$], a major product ion, with optimal collision energy for each compound. The PFB derivatives of DNA-adducts were detected with high sensitivity. The limit of detection was determined to be 200 fg (687 amol) for E-dGuo-PFB and 1 pg (2.48 fmol) for H-dGuo-PFB with a signal /noise ratio (S/N) of > 5:1.

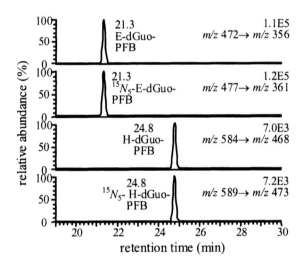

Figure 6. LC/MRM/MS analysis of E-dGuo, $^{15}N_5$-E-dGuo (internal standard), H-dGuo, and $^{15}N_5$-H-dGuoPFB derivatives (internal standard)

Extraction from Urine

The stable isotope internal standards (100 ng of each) were added to drug-free human urine (6 mL) from a laboratory volunteer and adjusted to pH 3. Liquid-liquid extraction using ethyl acetate was then performed to remove basic interferences from the urine sample, which provided a lower background level on LC/MRM//MS chromatogram than using the SPE method. E-dGuo and H-dGuo were stable in a weak-acidic condition (pH 3) and extracted into organic solvent with 55% and 75% of recoveries, respectively.

150

LC/MS Analysis of Normal Urine Sample

The organic layer containing the dGuo-adducts was dried and derivatized under mild conditions (room temperature, 2 h) with PFB-Br in the presence of the sterically hindered amine, DIPE. The derivatization mixture was evaporated to dryness so that the excess reagents were removed. The PFB-derivatized analytes were further purified by sequential use of a Si SPE column, followed by a C18 SPE column. The purified sample was analyzed by LC/MRM/MS using gradient system 1 as described above. The LC/MRM/MS chromatogram of a urine sample showed a peak at 24.8 min in the channel corresponding to m/z 584 $\rightarrow$ m/z 468 transition (Figure 7). This signal co-eluted with $^{15}N_5$-H-dGuo at 24.8 min in the m/z 589 $\rightarrow$ m/z 473 channel, so it was tentatively identified as H-dGuo at a level of approximately 0.1 ng/mL (Figure 7).

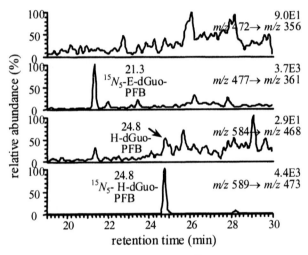

Figure 7. LC/MRM/MS analysis of E-dGuo, $^{15}N_5$-E-dGuo (internal standard), H-dGuo, and $^{15}N_5$-H-dGuo (internal standard) extracted from urine

However, it was difficult to precisely quantify the amount of H-dGuo because of the background interference. There was < 0.05 ng/mL of E-dGuo present in the urine sample.

In mouse skin carcinogenesis induced by PAH, increased etheno-adduct formation was observed as well as up-regulation of 12-LOX (8). This assay will make it possible to precisely establish the relative potency of each PAH and PAH o-quinone to stimulate etheno-adduct formation.

Summary

The quantitative analysis of DNA-adducts represents a formidable challenge because of their low abundance together with the ease with which artifacts can be generated. Methodology based on LC/MS offers the advantage that the intact adduct containing the sugar moiety can be analyzed. In the case of urinary adducts, co-eluting contaminants from the urine can cause substantial signal suppression in electrospray ionization so we have explored the use of APCI, which is less prone to such suppression effects. Initially, we explored the use of electron capture methodology coupled with normal phase chromatography. PFB derivatives of the etheno-dGuo adducts were prepared and found to give high intensity signals. Unfortunately, the background noise was unacceptable, particularly for urine samples. Therefore, the use of more conventional positive ion APCI methodology coupled with reversed phase chromatography was explored. PFB derivatization of E-dGuo and H-dGuo caused the expected shift in retention times on the reversed-phase column (Figures 4 and 5). This caused an increase in sensitivity because the solvents had a higher organic content and so APCI was more efficient. CID and MS/MS analysis revealed one major product ion for each analyte and its relevant stable isotope internal standard. This made it possible to conduct relatively sensitive MRM analysis of the two analytes and heavy isotope internal standards (Figure 6). Excellent recovery was obtained for the two internal standards from the urine samples as evidenced in Figure 7. Signals from E-dGuo and H-dGuo were a little obscured by co-eluting endogenous interfering substances that were extracted from the urine. However, the LC/MS assay will be useful for analysis of the two etheno-dGuo adducts in settings of oxidative stress when levels are expected to be much higher.

Acknowledgements

Supported by grants RO1-CA91016 and PO1-CA92537

References

1. Ames, B. N.; Shigenaga, M. K.; Hagen; T. M. *Proc. Natl. Acad. Sci. USA* **1993**, *90*, 7915-7922.
2. Penning, T. M.; Burczynski, M. E.; Hung, C-F.; McCoull, K. D.; Palackal, N. T.; Tsuruda, L. S. *Chem. Res. Toxicol.* **1999**, *12*, 1-18.
3. Dellinger, B.; Pryor, W. A.; Cueto, R.; Squadrito, G. L.; Hegde, V.; Deutsch, W. A. *Chem. Res. Toxicol.* **2001**, *14*, 1371-1377.

4. Porter, N. A.; Caldwell, S. E.; Mills, K. A. *Lipids* **1995**, *30*, 277-290.
5. Brash, A. R. *J. Biol. Chem.* **1999**, *274*, 23679-23682.
6. Laneuville, O.; Breuer, D. K.; Xu, N.; Huang, Z. H.; Gage, D. A.; Watson, J. T.; Lagarde, M.; DeWitt, D. L.; Smith, W. L. *J. Biol. Chem.* **1995**, *270*, 19330-19336.
7. Marnett, L. J.; DuBois, R. N. *Ann. Rev. Pharmacol. Toxicol.* **2002**, *42*, 55-80.
8. Nair, J.; Fürstenberger, G.; Burger, F.; Marks, F.; Bartsch, H. *Chem. Res. Toxicol.* **2000**, *13*, 703-709.
9. Burger, F.; Krieg, P.; Marks, F.; Furstenberger, G. *Biochem. J.* **2000**, *348*, 329-335.
10. Hamberg, M. *Arch. Biochem. Biophys.* **1998**, *349*, 376-380.
11. Marnett, L. J. *Carcinogenesis* **2000**, *21*, 361-370.
12. Lee, S. H.; Blair, I. A. *Chem. Res. Toxicol.* **2000**, *13*, 698-702.
13. Spiteller, P.; Kern, W.; Reiner, J.; Spiteller, G. *Biochim. Biophys. Acta* **2001**, *1531*, 188-208.
14. Schneider, C.; Tallman, K. A.; Porter, N. A.; Brash, A. R. *J. Biol. Chem.* **2001**, *276*, 20831-20838.
15. Lee, S. H.; Oe, T.; Blair, I. A. *Science* **2001**, *292*, 2083-2086.
16. Rindgen, D.; Nakajima, M.; Wehrli, S.; Xu, K.; Blair, I. A. *Chem. Res. Toxicol.* **1999**, *12*, 1195-1204.
17. Lee, S. H.; Rindgen, D.; Bible, R. A.; Hajdu, E.; Blair, I. A. *Chem. Res. Toxicol.* **2000**, *13*, 565-574.
18. Lin, J.; Fay, L. B.; Welti, D. H.; Blank, I. *Lipids* **2001**, *36*, 749-756.
19. Carvalho, V. M.; Asahara, F.; Di Mascio, P.; de Arruda Campos, I. P.; Cadet, J.; Medeiros, M. H. G. *Chem. Res. Toxicol.* **2000**, *13*, 397-405.
20. Loureiro, A. P. M.; Di Mascio, P.; Gomes, O. F.; Medeiros, M. H. G. *Chem. Res. Toxicol.* **2000**, *13*, 601-609.
21. Zamora, R.; Hidalgo, F. J. *Biochim. Biophys. Acta* **1995**, *1258*, 319-327.
22. Lee, S. H.; Oe, T.; Blair, I. A. *Chem. Res. Toxicol.* **2002**, *15*, 300-304.
23. Sattsangi, P. D.; Leonard, N. J.; Frihart, C. R. *J. Org. Chem.* **1977**, *42*, 3292-3296.

Aldo-Keto Reductases
and Exogenous Toxicants

Mycotoxins, Aldehydes, and Ketones

Chapter 11

Aflatoxin Aldehyde Reductases

Vincent P. Kelly[1], Tania O'Connor[1], Elizabeth M. Ellis[2], Linda S. Ireland[1],
Cara M. Slattery[1], Philip J. Sherratt[1], Dorothy H. Crouch[1],
Christophe Cavin[3], Benoît Schilter[3], Andrea Gallina[4], and John D. Hayes[1]

[1]Biomedical Research Centre, Ninewells Hospital and Medical School,
University of Dundee, Dundee DD1 9SY, Scotland, United Kingdom
[2]Bioscience and Pharmaceutical Sciences, University of Strathclyde,
Glasgow, G1 1XW, Scotland, United Kingdom
[3]Nestlé Research Centre, Vers-chez-les-Blanc, Lausanne 26, Switzerland
[4]University of Milan and National Research Council, Institute for Biochemical
and Evolutionary Genetics, 207 Abbiategrasso, Pavia, Italy

Aflatoxin B_1 is a potent hepatocarcinogen that is metabolized
by members of the aldo-keto reductase (AKR) superfamily.
The isoenzymes involved, called aflatoxin aldehyde
reductases, comprise the AKR7 family. They are dimeric
enzymes and uniquely catalyze the NADPH-dependent
reduction of the dialdehydic phenolate form of aflatoxin B_1-
8,9-dihydrodiol to a dialcohol. The facts that aflatoxin B_1-
dialdehyde readily forms protein adducts, and that certain of
the AKR7 isoenzymes that metabolize the dialdehyde are
markedly induced by cancer chemopreventive agents, suggests
that this family of reductases provides a measure of protection
against tumorigenesis by inhibiting the cytotoxic effects of
carcinogens. Besides aflatoxin B_1-dialdehyde, these enzymes
also exhibit activity towards dicarbonyls and presumably
contribute to the detoxification of such compounds. Members
of the AKR7 family are located either in the cytoplasm or are
associated with the Golgi apparatus. The latter enzymes
efficiently reduce succinic semialdehyde and may, because of
their subcellular location, make a major contribution to the
biosynthesis of γ-hydroxybutyrate, an endogenous
neuromodulatory molecule.

Introduction: The Distribution and Detoxification of Reactive Carbonyls

It is impossible to avoid exposure to chemicals containing reactive carbonyls as these compounds are widely distributed in the environment. They are many and varied, and include xenobiotics encountered in the air we breathe, the food we eat and in therapeutic agents. Carbonyls are also generated as a result of biotransformation of foreign compounds and during intermediate metabolism of biogenic amines, carbohydrates, vitamins and steroids. A further source of such compounds is from oxidation of protein and of DNA bases, as well as peroxidation of membrane lipids resulting from a range of toxic insults including iron overload, UV irradiation, redox-cycling drugs and inflammatory responses *(1-4)*.

Carbonyls pose a threat to life both because of their inherent reactivity towards cellular macromolecules and their relatively long half-life in the body. There is good evidence that carbonyls can form mutagenic adducts with DNA and can modify proteins *in vivo* *(1-7)*. Faced with this problem, various enzymes have evolved to detoxify aldehydes, ketones and quinones through catalysing either their metabolism in a NAD(P)H-dependent fashion or their conjugation with GSH. These include short-chain alcohol dehydrogenase, aldo-keto reductase (AKR), NAD(P)H:quinone oxidoreductase (NQO) and glutathione S-transferase (GST). The relative importance of these different enzyme systems in detoxifying carbonyls varies according to the chemical nature of the substance under consideration, and biological factors such as species, organ, sex and age of the organism.

Inducibility of the various reductases, as an adaptive response to electrophile or oxidative stress, is another variable that influences the contribution that different enzyme systems make to the detoxification of reactive carbonyls. The ability of cells to augment their capacity to detoxify foreign compounds in response to chemical insult, represents an important form of adaptation that has been recognized for many years *(8)*. For example, cytochrome P450 1A1 can be induced by planar aromatic hydrocarbons through the arylhydrocarbon receptor *(9-11)*, and NQO and GST can be induced by thiol-active agents through the bZIP transcription factor Nrf2 *(12)*.

Amongst the aldo-keto reductases, human dihydrodiol dehydrogenase AKR1C1 is inducible by metabolizable polycyclic aromatic hydrocarbons, β-naphthoflavone, *tert*-butylhydroquinone, dimethyl maleate, ethacrynic acid and organic isothiocyanates *(4,11,13)*, and rat AKR7A1 is inducible by a similar range of xenobiotics *(14,15)*. Thus, increased expression of AKR1C1 or AKR7A1 may profoundly alter the susceptibility of cells to reactive carbonyls. It is indeed interesting to note that AKR1C1 was found to be overexpressed in a human colon HT-29 cell line selected for resistance to ethacrynic acid *(16)*.

It is now apparent that AKR can play a pivotal role in the detoxification of foreign compounds in many mammalian species. However, this superfamily comprises numerous proteins, each of which possesses a broad and overlapping substrate specificity. It can therefore be difficult to ascertain the physiological significance of the varied reactions that individual AKR isoenzymes can catalyze *in vitro*. In this chapter, we describe the properties of the AKR7 family members and their apparently unique ability to detoxify a dialdehyde formed from aflatoxin B_1.

Discovery of Aflatoxin Aldehyde Reductase

The mycotoxin aflatoxin B_1 (AFB_1) is produced by *Aspergillus flavus*, a mould commonly found to contaminate cereal crops in humid regions of the world *(17)*. It is a potent liver carcinogen, but requires to be metabolized by hepatic cytochrome P450 to AFB_1-8,9-epoxide in order for it to exert its genotoxic effects. Exposure of Fischer 344 rats for 6 weeks to AFB_1 at 2 ppm in the diet is sufficient to produce hepatomas. However, the rats can be afforded complete protection against AFB_1 hepatocarcinogenesis if they are treated with the synthetic antioxidant ethoxyquin (at 0.5% (w/w) in the diet) prior to and concomitantly with the mycotoxin *(18,19)*. This type of anticarcinogen is called a "blocking" agent *(20)*. Resistance to the mycotoxin is associated with increased biliary excretion of an AFB_1-glutathione conjugate *(21)*, the fomation of which is catalyzed by the inducible class Alpha GST A5 subunit *(22,23)*; this GST polypeptide was called Yc_2 in an older nomenclature. Examination of AFB_1-8,9-epoxide metabolites generated *in vitro* by the hepatic cytosol prepared from rats fed on an ethoxyquin-containing diet, revealed that in addition to the glutathione conjugate, a dialcohol could also be detected *(24)*. In contrast with liver 100,000 g supernatants obtained from rats fed on a control diet, hepatic cytosol from rats fed on an ethoxyquin–containing diet catalyzed the production of > 5-fold more of both AFB_1 metabolites *(22,24,25)*. It was postulated by Judah *et al (24)* that the AFB_1-dialcohol was generated by reduction of AFB_1-dialdehyde, and that this carbonyl-containing metabolite had arisen as a result of rearrangement of the 8,9-dihydrodiol *(Figure 1)*.

The enzyme with aflatoxin aldehyde reductase activity (subsequently called AFAR) was first purified from the livers of rats fed an ethoxyquin-containing diet by sequential chromatography on DEAE-cellulose, CM-cellulose, hydroxyapatite and Protein-PAK Glass SP-8HR *(25)*. Although the purified enzyme was found to have activity towards the model AKR substrate 4-nitrobenzaldehyde, automated amino acid sequencing of CNBr peptides demonstrated AFAR is distinct from rat aldehyde reductase, aldose reductase and dihydrodiol dehydrogenase *(25,26)*. The notion that AFAR is a unique AKR

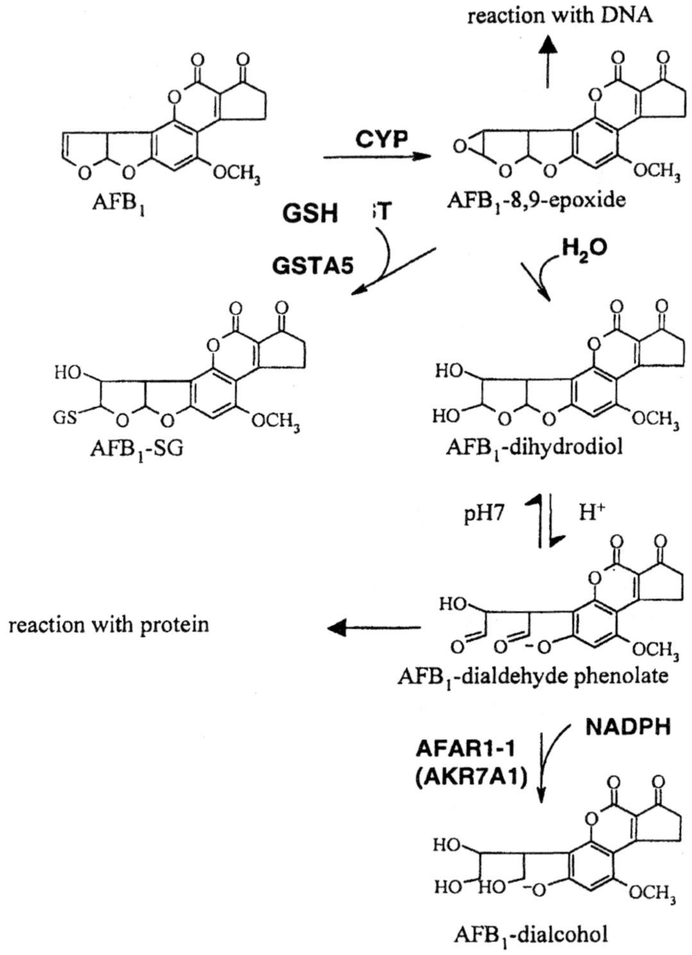

Figure 1. Major AFB₁ Detoxification Reactions in Rat Liver (adapted from (19)).

was supported by the Western blot data which revealed that the levels of the reductase are increased substantially in livers of ethoxyquin-treated rats when compared with control rat liver. Molecular cloning of the ethoxyquin-inducible AFAR provided definitive evidence that it is genetically separate from other reductases having activity towards 4-nitrobenzaldehyde. Indeed, it shares < 30% sequence identity with cDNAs encoding AKR1 proteins *(26)*. Thus, according to the nomenclature rules applied to this enzyme superfamily *(27,28)*, the inducible rat AFAR is the founding member of the AKR7 family and was designated AKR7A1 *(27,28)*. More recently, this rat enzyme has been shown by gel-filtration chromatography and X-ray crystallography to be dimeric *(29-31)*, and is more appropriately called rAFAR1-1, though in this article we call it AKR7A1-1 to conform to the nomenclature used throughout the monograph.

The available evidence shows that AFB_1-8,9-epoxide readily reacts with DNA *(17)*, whereas AFB_1-dialdehyde reacts with lysine residues in proteins *(17,32)*. Examination of the kinetic parameters of rat GST A5-5 in its ability to conjugate glutathione with epoxidated AFB_1, is consistent with the proposal that it protects against the genotoxic effects of the mycotoxin *(33)*. By similar reasoning, the *in vitro* kinetic properties of AKR7A1-1 are compatible with a protective role against the cytotoxic effects of AFB_1-dialdehyde *(34)*.

Substrate Specificity of AKR7A1-1 Indicates it is a Detoxication Enzyme

As indicated above, AKR7A1-1 is active towards AFB_1-dialdehyde and 4-nitrobenzaldehyde. It can however also reduce a number of aromatic aldehydes and aromatic dicarbonyls that serve as substrates for members of the AKR1A and AKR1B families *(35,36)*. For example, AKR7A1-1 is active towards 2-nitrobenzaldehyde, pyridine-2-aldehyde, 9,10-phenanthrenequinone, acenaphthenequinone, phenylglyoxal, isatin and 16-ketoestrone. The reductase is typically more active with diketones than towards aldehydes. A useful "diagnostic" substrate is 2-carboxybenzaldehyde *(37)* since this is not metabolized by AKR1A, AKR1B or AKR1C enzymes *(36)*. Amongst aliphatic aldehydes, AKR7A1-1 was found only to reduce succinic semialdehyde to γ-hydroxybutyrate (GHB), but is essentially inactive towards hexanal, propanal, formaldehyde, acetoacetic acid, acetone or methyl ethyl ketone *(35,36)*. Unlike AKR1C1 and AKR1C4, AKR7 isoenzymes appear to be poor at catalysing the reverse reaction. In particular, they are unable to catalyze the $NADP^+$-dependent oxidation of 1-acenaphthenol or cholic acid.

Identification of a Second Rat AKR7 Protein that acts as a GHB Synthase

Immunoblotting experiments led to the discovery of an additional member of the AKR7 family in the rat *(29)*. Subsequently, a combination of molecular cloning and protein chemistry approaches revealed that this second rat AKR7 polypeptide (rAFAR2, or AKR7A4) comprises 367 amino acids *(30)* whereas the AKR7A1 subunit is composed of 327 amino acids *(26)*. Figure 2 shows that the second rat AKR7 protein shares about 75% sequence identity with AKR7A1,

```
                                    S1
                    MSQARPATVLGAMEMGRRMD                             rAFAR1
                                 S  SC
MLRAVSRAVSRAAVRCAWRSGPSVARPLAMSRSPAPRAVSGAPLRPGTVLGTMEMGRRMD        rAFAR2
                                                   |
                                                   50

   H1          S2                        H2                   S3
VTSSSASVRAFLQRGHTEIDTAFVYANGQSETILGDLGLGLGRSGCKVKIATKAAPMFGK        rAFAR1
              B  SS S                                S  S SS
ASASAATVRAFLERGLNELDTAFMYCDGQSESILGSLGLGLGSGDCTVKIATKANPWDGK        rAFAR2
                                               |
                                              100

           H3            S4                       H4               S5
TLKPADVRFQLETSLKRLQCPRVDLFYLHFPDHGTPIEETLQACHQLHQEGKFVELGLSN        rAFAR1
                      SS                                         CC
SLKPDSVRSQLETSLKRLQCPRVDLFYLHAPDHGTPIVETLQACQQLHQEGKFVELGLSN        rAFAR2
                                  |
                                 150

         H5           S6              H6         S7
YVSWEVAEICTLCKKNGWIMPTVYQGMYNAITRQVETELFPCLRHFGLRFYAFNPLAGGL        rAFAR1
                                              C       CC C
YASWEVAEIYTLCKSNGWILPTVYQGMYNATTRQVETELLPCLRYFGLRFYAYNPLAGGL        rAFAR2
                                  |
                                 200
```

```
                          H7                  H8              250
LTGRYKYQDKGKNPESRFFGNPFSQLYMDRYWKEEHFNGIALVEKALKTTYGPTAPSMI    rAFAR1
      CC    S SS S
LTGKYRYEDKDGKQPEGRFFGNSWSETYRNRFWKEHHFEAIALVEKALKTTYGTDAPSMT   rAFAR2
      250                              300                300

   H9       S8      H10            H11
SAAVRWMYHHSQLKGTQGDAVILGMSSSLEQLEQNLALVEEGPLEPAVVDAFDQAWNLVAH   rAFAR1
              CCC C CC
SAALRWMYHHSQLQGTRGDAVILGMSSSLEQLEQNLAATEEGPLEPAVVEAFNQAWNVVAH   rAFAR2
                                         350

ECPNYFR    rAFAR1
       B
ECPNYFR    rAFAR2
```

Figure 2. Comparison between the primary structures of the AKR7A1 and AKR7A4 subunits. The residues in AKR7A1 associated with binding substrate are indicated by an "S", those associated with cofactor binding are indicated by a "C", and those that interact with both substrate and cofactor are indicated by a "B". Data taken from (30) and (31).

and that the extra 40 amino acids in rAFAR2 are found at the N-terminus as an Arg-rich domain that may form an amphipathic helix.

Like the AKR7A1 subunit, rAFAR2 (or AKR7A4) is found in rat liver as a homodimer. The fact that the two subunits can form a heterodimer means that quaternary structure may influence the function of the resulting AFAR1-2 isoenzyme. Anion-exchange chromatography allowed the rat hepatic enzymes to be resolved into rAFAR1-1, rAFAR1-2 and rAFAR2-2 (30); using the AKR nomenclature, these are referred to below as AKR7A1-1, AKR7A1-4 and AKR7A4-4. Examination of the catalytic properties of the reductases revealed that AKR7A4-4 is active towards many of the aromatic aldehydes and quinones that serve as substrates for AKR7A1-1. The AKR7A4 homodimer was found to have a higher K_m for some of these compounds, suggesting that it may not make as much of a contribution to detoxification of xenobiotics as AKR7A1-1. For example, with 2-carboxybenzaldehyde as substrate, AKR7A4-4 has a K_m value of ~10 µM, while AKR7A1-1 has a K_m of ~1 µM. By contrast with these observations, AKR7A4-4 was found to exhibit a significantly lower K_m towards succinic semialdehyde (i.e. 7 µM) than does AKR7A1-1 (i.e. 120 µM).

The AKR7A1-1 and AKR7A4-4 enzymes are located in different regions of the cell. Discontinuous gradient centrifugation and immunocytochemistry show that AKR7A1-1 is primarily a cytoplasmic reductase (30). By contrast, AKR7A4-4 is associated with the Golgi apparatus, and co-localizes with the Golgi marker G58 protein (30). It is not known whether AKR7A1-4 (ie rAFAR1-2) is primarily cytoplasmic or Golgi-associated. We conclude that AKR7A1-1 is a cytoplasmic detoxication enzyme, whereas AKR7A4-4 is involved in the synthesis of GHB through a Golgi-specific secretory pathway.

Identification of Two Aflatoxin Aldehyde Reductases in the Human

Following characterization of rat AFAR, two human reductases were cloned that can catalyze the reduction of AFB_1-dialdehyde. The first of these was described by Ireland et al (37) and is called AKR7A2. It has a primary structure that is more similar to AKR7A4 than to AKR7A1 (30). This human enzyme has K_m values of 10 µM and 40 µM for 9,10-phenanthrenequinone and 16-ketoestrone, respectively (Table 1). Significantly, it has a low K_m towards succinic semialdehyde, suggesting conservation in the function as a GHB synthase (37,38). Interestingly, antibodies that recognize the C-terminus of AKR7A2 (RW143) show the reductase to co-localize with the Golgi 58 kDa marker in human HepG2 cells (39). This suggests that, like AKR7A4, AKR7A2 contributes to GHB synthesis, and its association with the Golgi apparatus facilitates secretion of the hydroxylated product. The full-length cDNA for AKR7A2 encodes a polypeptide of 359 amino acids that, like AKR7A4, also contains an additional N-terminal domain (30). It does however appear that, at least in the case of AKR7A2 isolated from human liver, translation can be initiated from the second in-frame ATG codon resulting in production of a polypeptide of 330 amino acids *(37)*. A number of expressed sequence tags have been obtained from various human tissues/tissue lines including skin, muscle and adenocarcinomas of colon that contain the first in-frame ATG codon. It remains to be established whether AKR7A2 protein in the human adrenal gland comprises 359 or 330 amino acids. It is also not known if there are specific cellular mechanisms that control whether translation is initiated from either the first or second in-frame ATG in the AKR7A2 mRNA.

The second human AFAR protein, AKR7A3, was described by Knight *et al (40)* and has a primary structure that is more closely related to AKR7A1 than to AKR7A4 *(30)*. Like AKR7A1, AKR7A3 is composed of 327 amino acids. AKR7A3 has a higher activity towards AFB_1-dialdehyde than does AKR7A2 (Dr. F. P. Guengerich and Dr. T. R. Sutter, personal communication). Although Knight *et al (40)* have shown AKR7A3 is active towards 4-nitrobenzaldehyde and 9,10-phenanthrenequinone its catalytic specificity remains to be thoroughly characterized.

Division of the AKR7 Family into Two Classes

Characterization of the rat and human AKR7 enzymes suggests that they can be divided into two groups. Specifically, AKR7A1 and AKR7A3 appear to be more closely related to each other than to AKR7A4 and AKR7A2 *(30)*. The full-length cDNAs encoding the latter pair of reductases are significantly longer than those encoding the former pair of enzymes. Recently, a mouse reductase designated AKR7A5 has been cloned and characterized as a bacterially-expressed protein *(41)*. This murine reductase has a low K_m value for succinic semialdehyde *(41)* and can be placed in the same subclass as AKR7A4 and AKR7A2 *(30)*. It remains to be established whether AKR7A5 has an additional N-terminal domain and can associate with the Golgi apparatus.

Regulation of AKR7 Proteins

The original discovery of AKR7A1 as an ethoxyquin-inducible protein raised questions about the mechanisms responsible for its regulation. A significant number of inducers for this protein have now been identified. These include butylated hydroxyanisole, benzylisothiocyanate, coffee diterpenes (cafestol and kahweol), coumarin, diallyl disulfide, diethylmaleate, dithiolethione, oltipraz, *trans*-stilbene oxide and β-naphthoflavone *(14,15,42-45)*. Some of the compounds that induce AKR7A1 are known cancer chemopreventive agents. Other compounds are phytochemicals found in Cruciferous vegetables, *Leguminosae* species, Garlic and Coffee beans, all of which appear to have anticarcinogenic actions. The structures of these inducing agents are shown in Figure 3. It is interesting to note that some of the compounds

Table 1. Substrate specificity of rat (AKR7A1 and AKR7A4) and human (AKR7A2 and AKR7A3) reductases. The data are taken from (30,34,37).

substrate		protein	K_m (μM)	k_{cat} (min^{-1})	k_{cat}/K_m (min^{-1}M^{-1})
aflatoxin B$_1$-dialdehyde	MA1	AKR7A1	—	—	2.5x10^4
		AKR7A3	36 ±6	2 ±0.2	5.6x10^4
	MA2	AKR7A1	109 ±48	12 ±3	1.1x10^5
		AKR7A3	127 ±23	18 ±2	1.4x10^5
succinic semialdehyde		AKR7A1	184 ±33	69 ±8	3.8x10^5
		AKR7A4	6.4 ±0.6	76 ±1	1.2x10^7
		AKR7A2	15.4 ±2	92 ±3	6.0x10^6
2-carboxy-benzaldehyde		AKR7A1	0.76 ±0.3	74 ±4	9.7x10^7
		AKR7A4	9.7 ±0.2	82 ±1	8.5x10^6
		AKR7A2	15.4 ±2	101 ±5	5.1x10^6
2-nitro-benzaldehyde		AKR7A1	1100 ±21	490 ±5	4.5x10^5
		AKR7A2	2430 ±320	109 ±8	4.5x10^4
4-nitro-benzaldehyde		AKR7A1	3600 ±1000	247 ±40	6.9x10^4
		AKR7A2	6100 ±900	94 ±10	1.5x10^4

	9,10-phenathrene quinone	AKR7A1	9.6 ±2.7	731 ±66	7.6x10^7
		AKR7A2	9.0 ±3.5	325 ±39	3.6x10^7
	16-oxo-oestrone	AKR7A1	24 ±5	518 ±42	2.2x10^7
		AKR7A2	40 ±6	142 ±10	3.6x10^6
	isatin	AKR7A1	105 ±87	87 ±29	8.3x10^5
		AKR7A2	248 ±31	217 ±16	8.8x10^5
	NADPH	AKR7A1 (with SSA)	0.32 ±0.14	91 ±10	2.8x10^8
		AKR7A4 (with 2-CBA)	1.46 ±0.1	86 ±8	5.9x10^7
		AKR7A2 (with SSA)	0.75 ±0.3	108 ±9	1.4x10^8

166

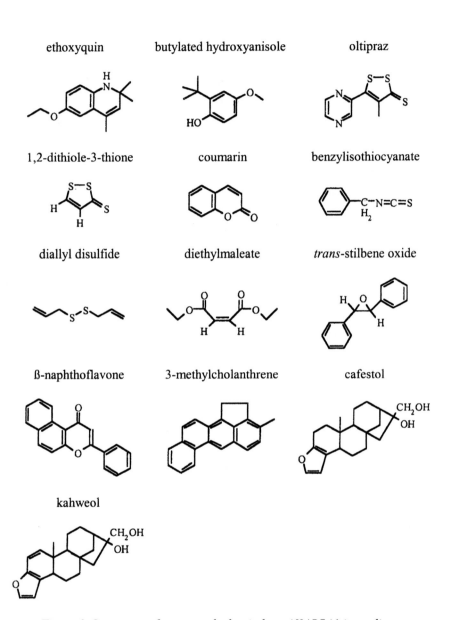

Figure 3. Structures of compounds that induce AKAR7A1 in rat liver.

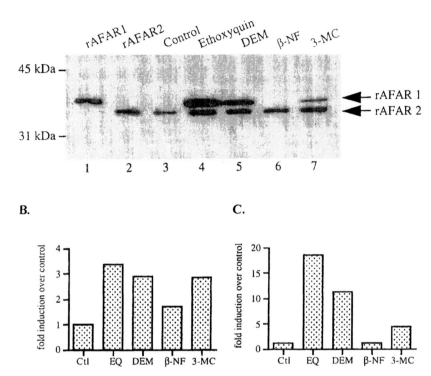

Figure 4. Regulation of hepatic AKR7A4 by xenobiotics. Panel A shows a Western blot of liver cytosol from rats fed on a control diet (lane 3), a diet containing 0.5% ethoxyquin (lane 4), or a diet containing 0.5% diethylmaleate (lane 5). Other rats were administered i.p. β-naphthoflavone at 200 mg/kg (lane 6), or were administered i.p. 3-methylcholanthrene at 200 mg/kg (lane 7). After SDS/PAGE and transfer to immobilon P, the blot was probed with an anti-AKR7A2 serum which cross-reacts with both AKR7A1 and AKR7A4. Bacterially-expressed, AKR7A1 (i.e. rAFAR1) and AKR7A4 (i.e. rAFAR2) were loaded in lanes 1 and 2, respectively, as positive controls. Panel B shows a quantification of the AKR7A4 subunit, by phosphoimaging the ECL-developed blot. Panel C shows estimation of the amount of AKR7A1 on the blot.

that induce rat AKR7A1 also induce human AKR1C1 *(10,11,13)*, suggesting that the two genes may be subject to common regulatory mechanisms.

In addition to induction by xenobiotics, expression of AKR7A1 is increased in the livers of selenium-deficient rats *(46)*. In this instance, the overexpression of the reductase has been attributed to accumulation of intracellular H_2O_2 caused by loss of the selenium-dependent glutathione peroxidase. During hepatocarcinogenesis, AKR7A1 is overexpressed in livers containing pre-neoplastic nodules as well as in hepatomas *(19,25)*.

The range and types of xenobiotics that induce AKR7A1 is indicative of the presence of an antioxidant responsive element (ARE, $5'-^A/_G$GTGACNNNGC-3') in the promoter of the gene *(47)*. The 5'-upstream region of *AKR7A1* has been isolated and found to contain several potential ARE enhancers. In particular, multiple copies of the motif 5'-GTGAG-3' were found around nucleotide -390 *(48)*. Although no perfect AREs have been identified in the *AFAR1* promoter, related sequences include 5'-AATGATTCAGCA-3' and 5'-CCTGAGTGAGCG-3' *(48)*. Regulation of *AKR7A1* resembles rat *GSTA5*, and this gene also possesses an ARE *(49)*.

The AKR7A4 subunit in rat prostate is inducible by androgens *(50)*. While little is known about regulation of this reductase by xenobiotics, Figure 4 shows it is modestly inducible by ethoxyquin, diethylmaleate, β-naphthoflavone and 3-methylcholanthrene. To date, there is no evidence that the human or mouse AKR7 proteins are inducible.

Conclusions

This article outlines the discovery of the aflatoxin aldehyde reductases. Although it appears that they serve to protect against cytotoxicity caused by the mycotoxin, their other biological functions are poorly understood. It is not known why AKR7A4 and AKR7A2 associate with the Golgi apparatus. The mechanisms that control this process have yet to be described.

Acknowledgements

We acknowledge financial support from the AICR (98-023), BBSRC (94/DO8200), MRC (G9322073PA) and Wellcome Trust.

References

1. Stadtman, E. R.; Berlett, B. S. *Chem. Res. Toxicol.* **1997**, *10*, 485-494.
2. Wiseman, H.; Halliwell, B. *Biochem. J.* **1996**, *313*, 17-29.
3. Henle, E. S.; Linn, S. *J. Biol. Chem.* **1997**, *272*, 19095-19098.

4. Burczysnki, M. E.; Lin, H.-K.; Penning, T. M. *Cancer Res.* **1999**, *59*, 607.
5. Esterbauer, H.; Schaur, R. J.; Zollner, H. *Free Radical Biol. Med.*, **1991**, *11*, 81-128.
6. Chaudhary, A. K.; Nokubo, M.; Reddy, G. R.; Yeola, S. N.; Morrow, J. D.; Blair, I. A.; Marnett, L.J. *Science* **1994**, *265*, 1580-1582.
7. Dedon, P. C.; Plastaras, J. P.; Rouzer, C. A.; Marnett, L. J. *Proc. Natl. Acad. Sci. USA* **1998**, *95*, 11113-11116.
8. Parkinson, A. In *Casarett & Doull's Toxicology. The Basic Science of Poisons (fifth edition) Klaassen, C.D., Ed.; McGraw-Hill, New York 1996;* pp 113-186.
9. Hankinson, O. *Annu. Rev. Pharmacol. Toxicol.* **1995**, *35*, 307-340.
10. Burczynski, M. E.; Penning, T. M. *Cancer Res.* **2000**, *60*, 908-915.
11. Bonnesen, C.; Eggleston, I. M.; Hayes, J. D. *Cancer Res.* **2001**, *61*, 6120-6132.
12. McMahon, M.; Itoh, K.; Yamamoto, M.; Chanas, S. A.; Henderson, C. J.; McLellan, L. I.; Wolf, C. R.; Cavin, C.; Hayes, J. D. *Cancer Res.* **2001**, *61*, 3299-3307.
13. Ciaccio, P. J.; Jaiswal, A. K.; Tew, K. D. *J. Biol. Chem.* **1994**, *269*, 15558-15562.
14. Ellis, E. M.; Judah, D. J.; Neal, G. E.; O'Connor, T.; Hayes, J. D. *Cancer Res.* **1996**, *56*, 2758-2766.
15. Kelly, V. P.; Ellis, E. M.; Manson, M. M.; Chanas, S. A.; Moffat, G. J.; McLeod, R.; Judah, D. J.; Neal, G. E.; Hayes, J. D. *Cancer Res.* **2000**, *60*, 957-969.
16. Ciacco, P. J.; Stuart, J. E.; Tew, K. D. *Mol. Pharmacol.* **1993**, *43*, 845-853.
17. Eaton, D. L.; Gallagher, E. P. *Annu. Rev. Pharmacol. Toxicol.* **1994**, *34*, 135.
18. Cabral, J. R. P.; Neal, G. E. *Cancer Lett.* **1983**, *19*, 125-132.
19. Hayes, J. D.; McLeod, R.; Ellis, E. M.; Pulford, D. J.; Ireland, L. S.; McLellan, L. I.; Judah, D. J.; Manson, M. M.; Neal, G. E. In *Principles of Chemoprevention. International Agency for Research (IARC) Scientific Publications* **1996**, *no. 139*, pp 175-187.
20. Wattenberg, L. W. *Cancer Res.* **1985**, *45*, 1-8.
21. Kensler, T. W.; Egner, P. A.; Davidson, N. E.; Roebuck, B. D.; Pikul, A.; Groopman, J. D. *Cancer Res.* **1986**, *46*, 3924-3931.
22. Hayes, J. D.; Judah, D. J.; McLellan, L. I.; Kerr, L. A.; Peacock, S. D.; Neal, G.E. *Biochem. J.* **1991**, *279*, 385-398.
23. Hayes, J. D.; Nguyen, T.; Judah, D. J.; Petersson, D. G.; Neal, G. E. *J. Biol. Chem.* **1994**, *269*, 20707-20717.
24. Judah, D. J.; Hayes, J. D.; Yang, J.-C.; Lian, L.-Y.; Roberts, G. C. K.; Farmer, P. B.; Lamb, J. H.; Neal, G. E. *Biochem. J.* **1993**, *292*, 13-18.
25. Hayes, J. D.; Judah, D. J., Neal, G. E. *Cancer Res.* **1993**, *53*, 3887-3894.
26. Ellis, E. M.; Judah, D. J.; Neal, G. E.; Hayes, J. D. *Proc. Natl. Acad. Sci. USA* **1993**, 90, 10350-10354.

27.   Jez, J. M.; Flynn, T. G.; Penning, T. M. *Biochem. Pharmacol.* **1997**, *54*, 639-647.

28.   Jez, J. M.; Penning, T. M. *Chemico.-Biol. Interact.* **2001**, *130-132*, 499-525.

29.   Kelly, V. P.; Ireland, L. S.; Ellis, E. M.; Hayes, J. D. *Biochem. J.* **2000**, *348*, 389-400.

30.   Kelly, V. P.; Sherratt, P. J.; Crouch, D. H.; Hayes, J. D. *Biochem. J.* **2002**, *366*, 847-861.

31.   Kozma, E.; Brown, E.; Ellis, E. M.; Lapthorn, A. J. *J. Biol. Chem.* **2002**, *277*, 16285-16293.

32.   Guengerich, F. P.; Arneson, K. O.; Williams, K. M.; Deng, Z.; Harris, T. M. *Chem. Res. Toxicol.* **2002**, *15*, 780-792.

33.   Johnson, W. W.; Ueng, Y.-F.; Widersten, M.; Mannervik, B.; Hayes, J. D.; Sherratt, P.J.; Ketterer, B.; Guengerich, F.P. *Biochemistry* **1997**, *36*, 3056-3060.

34.   Guengerich, F. P.; Cai, H.; McMahon, M.; Hayes, J. D.; Sutter, T. R.; Groopman, J. D.; Deng, Z.; Harris, T. M. *Chem. Res. Toxicol.* **2001**, *14*, 727-737.

35.   Ellis, E. M.; Hayes, J. D. *Biochem. J.* **1995**, *312*, 535-541.

36.   O'Connor, T.; Ireland, L. S.; Harrison, D. J.; Hayes, J. D. *Biochem. J.* **1999**, *343*, 487-504.

37.   Ireland, L. S.; Harrison, D. J.; Neal, G. E.; Hayes, J. D. *Biochem. J.* **1998**, *332*, 21-34.

38.   Schaller, M.; Schaffhauser, M.; Sans, N.; Wermuth, B. *Eur. J. Biochem.* **1999**, *265*, 1056-1060.

39.   Gallina, A.; Milanesi, G.; Hayes, J. D. unpublished results.

40.   Knight, L. P.; Primiano, T.; Groopman, J. D.; Kensler, T. W.; Sutter, T. R. *Carcinogenesis* **1999**, *20*, 1215-1223.

41.   Hinshelwood, A.; McGarvie, G.; Ellis, E. *FEBS Lett.* **2002**, *523*, 213-218.

42.   Anderson, L.; Steele, V.K.; Kelloff, G.J.; Sharma, S. *J. Cell. Biochem.*, suppl. **1995**, *22*, 108-116.

43.   Primiano, T.; Gastel, J.A.; Kensler, T. W.; Sutter, T. R. *Carcinogenesis* **1996**, *17*, 2297-2203.

44.   Guyonnet, D.; Belloir, C.; Suschetet, M; Siess, M.-H.; Le Bon, A.-M. *Carcinogenesis* **2002**, *23*, 1335-1341.

45.   Cavin, C.; Holzhaeuser, D.; Bezencon, C.; Guignard, G.; Schilter, B. *Proc. Am. Ass. Cancer Res.* **2002**, poster 3430.

46.   McLeod, R.; Ellis, E. M.; Arthur, J. R.; Neal, G. E.; Judah, D. J.; Manson, M. M.; Hayes, J. D. *Cancer Res.* **1997**, *57*, 4257-4266.

47.   Hayes, J. D.; McMahon, M. *Cancer Lett.* **2001**, *174*, 103-113.

48.   Ellis, E. M.; Slattery, C. M.; Hayes, J. D. unpublished results

49.   Pulford, D. J.; Hayes, J. D. *Biochem. J.* **1996**, *318*, 75-84.

50.   Nishi, N.; Shoji, H.; Miyanaka, H.; Nakamura, T. *Endocrinology* **2000**, *141*, 3194-3199.

Chapter 12

Competing Reactions of Aflatoxin B$_1$ Dialdehyde: Enzymatic Reduction versus Adduction with Lysine

F. Peter Guengerich[1], Kevin M. Williams[1], Thomas R. Sutter[2],
John D. Hayes[3], William W. Johnson[1], Kyle O. Arneson[1], Markus Voehler[1],
Zhenwu Deng[1], and Thomas M. Harris[1]

[1]Departments of Biochemistry and Chemistry and Center in Molecular Toxicology,
Vanderbilt University, Nashville, TN 37232
[2]W. Harry Feinstone Center for Genomic Research, University of Memphis,
Memphis, TN 38152
[3]Biochemical Research Centre, Ninewells Hospital and Medical School,
University of Dundee, Dundee DD1 9SY, Scotland, United Kingdom

The mycotoxin aflatoxin (AF)B$_1$ is a strong human hepatocarcinogen and can also cause acute toxicity. A long-term project has resulted in the characterization of the enzymatic formation of and reactions of AFB$_1$ exo-8,9-epoxide, a product that plays a central role in the metabolism. The dihydrodiol formed in the hydrolysis of AFB$_1$ 8,9-epoxide is in equilibrium with its ring-opened dialdehyde form. The dialdehyde can be reduced to unreactive mono- and dialcohol derivatives by human AFB$_1$ aldehyde reductases (e.g. AKR7A2, AKR7A3) or react with protein lysines to form oxopyrrole products. Kinetic parameters for these and other AFB$_1$-related reactions have been measured and provide insight into the respective pathways.

171

Aflatoxin (AF)B$_1$ was originally discovered because of the acute death of turkeys eating mold-contaminated feed. Subsequently, this mycotoxin was shown to be hepatocarcinogenic and a serious health risk in parts of Africa and Asia. The history, occurrence, and epidemiology have been reviewed elsewhere (*1*). In the 1970s, evidence was presented that metabolism plays a major role in the toxicities of AFB$_1$, as well as its detoxication. A general scheme developed by the early 1980s is shown in Figure 1.

The central role of the epoxide was inferred from some of the products (*1,3*) and confirmed following the synthesis in 1988 (*4*). The half-life of the epoxide was measured and found to be 1 s at 23 °C at neutral pH (*5*). The resulting dihydrodiol is in equilibrium with the dialdehyde, with an effective pK_a of 8.2. A problem in the analysis of reactions of AFB$_1$ dialdehyde is that most of the experiments in the earlier literature were done under conditions in which the dihydrodiol predominated. Experiments were done with pre-formed dialdehyde under conditions in which the rearrangement to dihydrodiol was limited, based upon our previous kinetic studies (*5*).

Materials and Methods

AFB$_1$ *exo*-8,9-epoxide was prepared and crystallized as described (*6*). The AFAR enzymes were prepared as described (*2*). Details of the conditions for incubations, HPLC, UV and NMR spectroscopy, and mass spectrometry are described elsewhere (*2,7,8*).

Results and Discussion

The general approach used in incubations with AFB$_1$ dialdehyde was to prepare the compound by hydrolysis of AFB$_1$ *exo*-8,9-epoxide in 20 mM sodium 2-(*N*-cyclohexylamino) ethanesulfonate buffer (pH 10.0). This material, freshly prepared, was stored on ice the day of use. Aliquots were added to other reactants (e.g. AFAR) in neutral pH buffer to begin reactions. Reaction times were minimized to prevent the rearrangement of AFB$_1$ dialdehyde to the dihydrodiol (*2,5,7,8*).

Chemistry of Reduction of AFB$_1$ Dialdehyde by AFAR

Reduction of AFB$_1$ dialdehyde with rat AFAR (AKR7A1) and NADPH for 60 s yielded the products shown in the HPLC chromatogram of Figure 2, with UV spectra shown in the inset. Mass spectral analysis indicated that peaks 1 and 2 were monoaldehydes and peak 3 was the subsequent reduction product, AFB$_1$

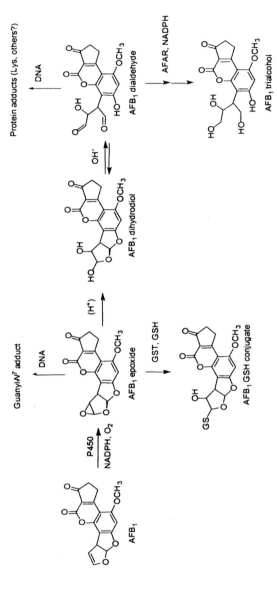

Figure 1. Major steps in AFB₁ metabolism (adapted from (2))

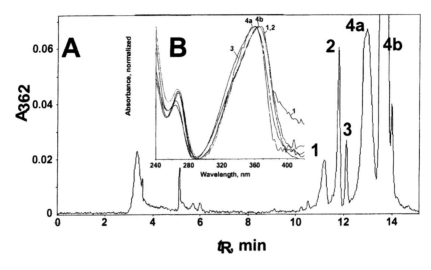

Figure 2. HPLC chromatograms (A) and UV spectra (B) of reduction products of AFB₁ dialdehyde obtained with rat AFAR (AKR7A1).
The λ_max values (recorded on-line, inset B) for the peaks were, respectively, 1(AFB₁ monoaldehyde 1): 336 nm, 2 (AFB₁ monoaldehyde 2): 367 nm, 3 (AFB₁ triol): 361 nm, 4a (AFB₁ dialdehyde, form 1): 360 nm, and 4b (AFB₁ dialdehyde, form 2): 365 nm. The respective MH⁺ (protonated molecular ion) peaks (electrospray mass spectrometry) were (m/z) 1: 349.2, 2: 349.2, 3: 351.3, and 4a and 4b: 347.
(Reproduced from reference 2. Copyright 2001 American Chemical Society.)

triol (Figure 1). Peaks **4a** and **b** were equilibrating forms of the dialdehyde substrate. The same products were formed by brief limited reduction of AFB₁ dialdehyde with NaBH₄. The products were not formed in the enzymatic reaction when the dialdehyde had been standing at pH 7 prior to the addition of AFAR.

The identities of the individual monoaldehydes were determined in the following manner. Peaks **1** and **2** were separated from a short incubation with rat AKR7A1 and then reduced with NaBD₄ (Figure 3). The resulting triols were cleaved with NaIO₄, and the residual deuterium was analyzed by mass spectrometry, thus establishing the structures.

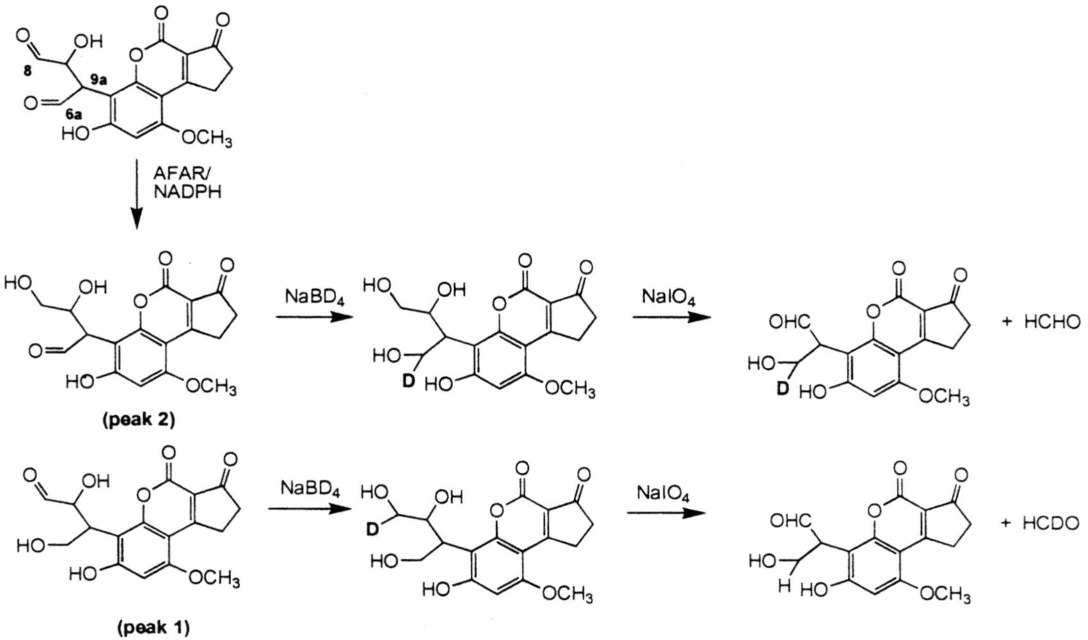

Figure 3. Analysis of monoaldehyde reduction products of AFB_1 dialdehyde using $NaIO_4$ (2)

AFAR Reduction Rates

The reduction of AFB_1 dialdehyde (by rat AKR7A1) to monoaldehyde isomer **2** was measured by HPLC methods and found to be much faster than to monoaldehyde **1** (Figure 4). The parameters $k_{cat} = 12$ min^{-1} and $K_m = 110$ μM (AFB_1 dialdehyde) were estimated for the formation of monoaldehyde **2** (*2*). Subsequent work with human AKR7A2 yielded $k_{cat} = 20$ ($\pm$ 5) min^{-1} and $K_m = 610$ ($\pm$ 190) μM; the values with human AKR7A3 were $k_{cat} = 4.3$ ($\pm$ 0.4) min^{-1} and 51 ($\pm$ 11) μM.

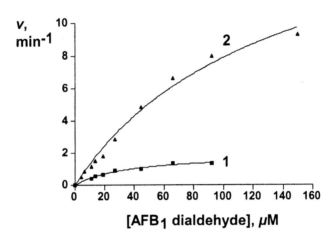

Figure 4. Rates of reduction of AFB_1 dialdehyde to monoaldehyde by rat AKR7A1 (2). For monoaldehyde 2 (▲) the parameters $k_{cat} = 12$ min^{-1} and K_m 110 μM were estimated. For monoaldehyde 1 (■) the ratio $k_{cat}/K_m = 0.025$ min^{-1} μM^{-1} was estimated because of the lack of saturation.

Rates of NADPH oxidation associated with AFB_1 dialdehyde reduction could not be directly measured by ΔA_{340} measurements because of the increase in absorbance associated with AFB_1 dialdehyde rearrangement to the dihydrodiol. Fitting of the ΔA_{340} data (Figure 5) to a basic kinetic scheme that described the simultaneous processes yielded an estimate of the enzyme efficiency an order of magnitude higher (*2*). Presumably this increased efficiency applies also to the human AKR enzymes.

Chemistry of Amine Adducts Formed from AFB_1 Dialdehyde

Previous work indicated that the major protein adducts formed from AFB_1 were lysine adducts (*9*). However, several oxopyrrole derivatives are possible (Figure 6).

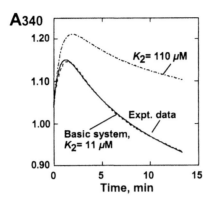

Figure 5. Fitting of experimental plot of A_{340} to a reaction mechanism with k_2, the (rat) AFAR-AFB$_1$ dialdehyde dissociation constant, set at 11 μM instead of the K_m of 110 μM estimated in the experiment (from (2)).

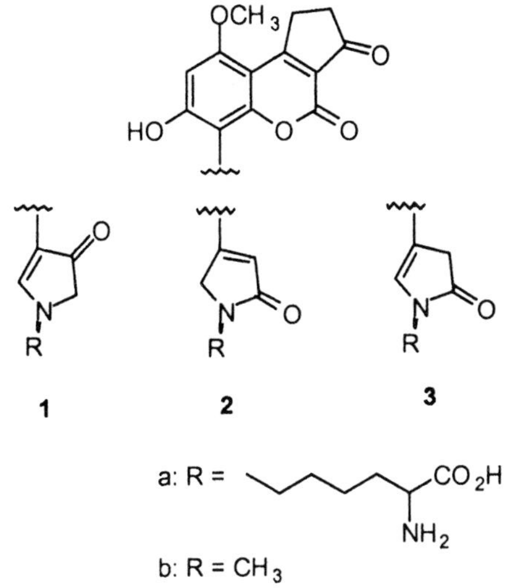

Figure 6. Possible structures of amine adducts formed from AFB$_1$ dialdehyde (8)

Both **1** and **2** have been reported in the literature (*9,10*), and the assignments depend upon comparisons of UV spectra and NMR chemical shifts.

As a model, we prepared the methylamine adduct of AFB$_1$ dialdehyde, which had the UV and fluorescence spectra of the lysine adduct (*8*). NMR 2-dimensional nuclear Overhauser spectra are only consistent with structure **2b** (Figure 6), and we infer that the lysine adduct has structure **2a**.

The establishment of the structure is of significance in the context of the mechanism of the reaction of AFB$_1$ dialdehyde with lysine. In Figure 7, path **a** would involve a postulated Amodori rearrangement (*9*). Path **b**, which yields the assigned structure, is based on dehydration and pyrrole rearrangement (*8,10*).

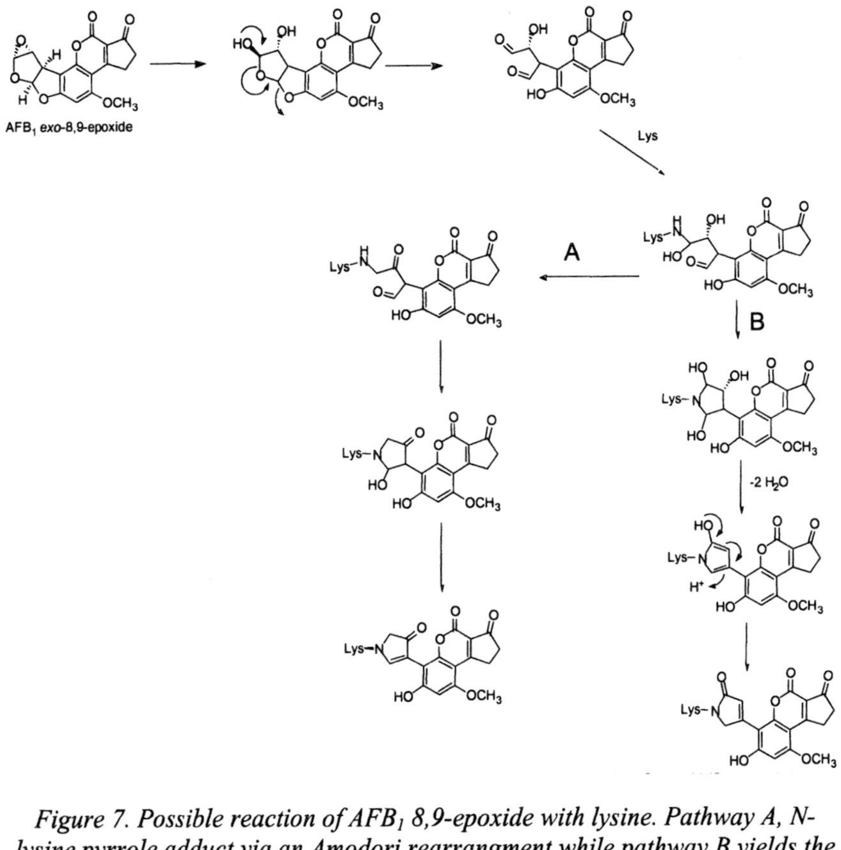

Figure 7. Possible reaction of AFB$_1$ 8,9-epoxide with lysine. Pathway A, N-lysine pyrrole adduct via an Amodori rearrangment while pathway B yields the same structure via dehydration and rearrangement.

One of the lysine derivatives of AFB_1 that was given some consideration was AFB_{2a}, which is readily formed by acid-catalyzed hydration (*1*). The extent of the presence of AFB_{2a} in biologically-relevant samples is not clear (*1*). The elucidation of the chemistry of AFB_1 dialdehyde with amines would suggest the reaction pathways shown in Figure 8, yielding a pyrrole derivative, which might or might not be stable to further reaction.

Kinetics of Reaction of AFB_1 Dialdehyde with Lysines

AFB_1 dialdehyde was reacted with bovine serum albumin; proteolysis of the products and HPLC-fluorescence were used in analysis. The process could be characterized by $K_d = 1.5$ mM and $k = 0.033$ min^{-1} for the reaction (*7*). The reaction of N^2-acetyllysine with AFB_1 dialdehyde could be characterized by a second-order rate constant of 2.6×10^3 M^{-1} min^{-1}, followed by a "rearrangement" rate of 7.6 min^{-1}. Thus, the reaction of AFB_1 dialdehyde with a ("average") lysine in albumin appears to be attenuated by steric restriction or protonation of protein lysines.

Mass spectrometry was used to identify the residues Lys455 and Lys548 as preferential sites of modification (*7*).

AFB_1 *exo*-8,9-epoxide can react directly with (N^2-acetyl)lysine, as demonstrated by the attenuated dihydrodiol formation with high concentrations of this amine (*7*). The inhibition was more pronounced at pH 9.5 (than at 7.2), as expected. This result provides further evidence for the S_N2 nature of the epoxide hydrolysis. The kinetics indicate that the direct reaction of the epoxide could contribute to the formation of lysine adducts, although the reaction with the dialdehyde is probably more favorable (*7*).

Conclusions

The experiments on the reactions of AFB_1 dialdehyde can be coupled with previous work to fill in the kinetic parameters in Figure 9 (*2,5,7,11-13*).

The uncertainty in the rates of reactions with AFB_1 dialdehyde is due to uncertainty about the appropriateness of the raw kinetic parameters (in the case of the AFAR reaction) and the variability between N^2-acetyllysine and albumin lysine (in the case of the protein reaction).

The scheme provides a context for consideration of the roles of individual reactions in the overall disposition of AFB_1. The rapid rate of hydrolysis of AFB_1 *exo*-8,9-epoxide argues against a role for the enzyme epoxide hydrolase (*14*). The glutathione transferase results are supported by the absence of the glutathione conjugate in hepatocytes from individuals devoid of the enzyme (*15*). The short half-life of the epoxide also argues that transport out of hepatocytes should be unlikely, although this point has not been directly examined. The reaction rates of some of the AFAR enzymes suggest that they are capable of contributing to the detoxication of AFB1 dialdehyde and blocking its reaction with protein lysines, thus serving in a detoxication role.

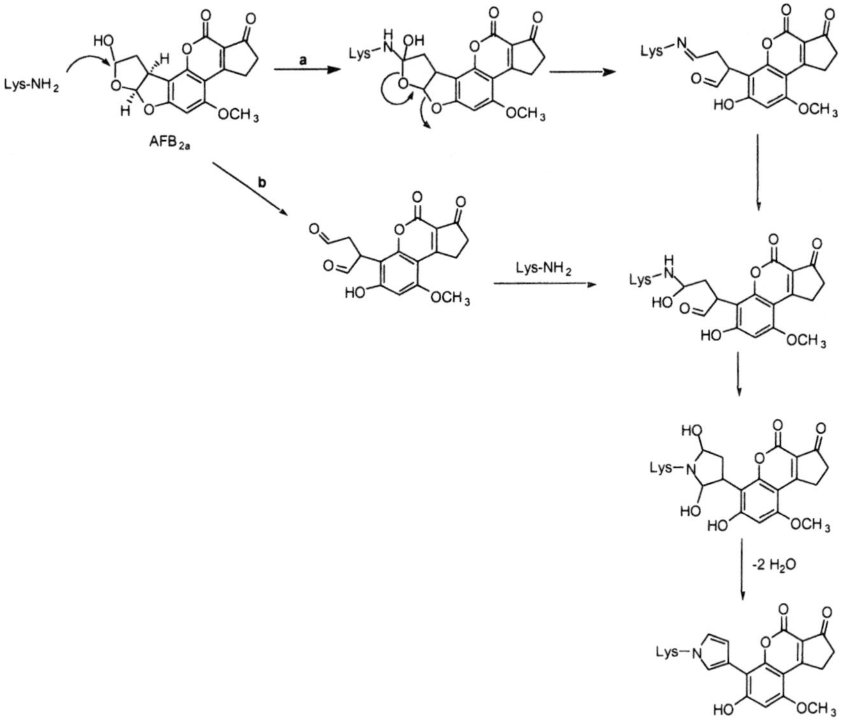

Figure 8. Putative reaction of AFB₂ₐ with lysine. (a) direct attack of lysine on
AFB₂ₐ. (b) ring opening of AFB₂ₐ followed by reaction of the resulting
dialdehyde with lysine to form a pyrrole.

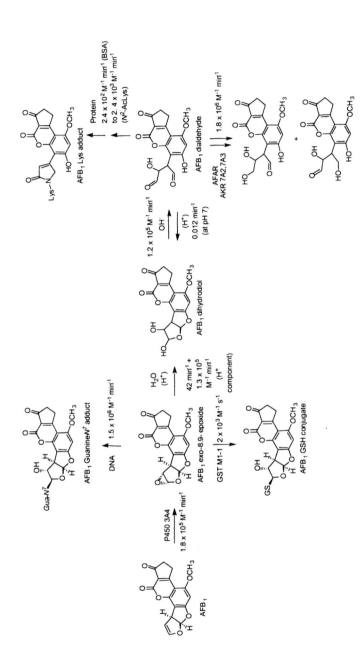

Figure 9. Summary of major steps in human AFB_1 metabolism with estimated rates (adapted from (7))

References

1. *Chemical Carcinogens*; Searle, C. E., Ed.; American Chemical Society: Washington, DC, 1984, pp 945-1136.
2. Guengerich, F. P.; Cai, H.; McMahon, M.; Hayes, J. D.; Sutter, T. R.; Groopman, J. D.; Deng, Z.; Harris, T. M. *Chem. Res. Toxicol.* **2001**, *14*, 727-737.
3. Essigmann, J. M.; Croy, R. G.; Nadzan, A. M.; Busby, W. F., Jr.; Reinhold, V. N.; Büchi, G.; Wogan, G. N. *Proc. Natl. Acad. Sci. USA* **1977**, *74*, 1870-1874.
4. Baertschi, S. W.; Raney, K. D.; Stone, M. P.; Harris, T. M. *J. Am. Chem. Soc.* **1988**, *110*, 7929-7931.
5. Johnson, W. W.; Harris, T. M.; Guengerich, F. P. *J. Am. Chem. Soc.* **1996**, *118*, 8213-8220.
6. Iyer, R. S.; Harris, T. M. *Chem. Res. Toxicol.* **1993**, *6*, 313-316.
7. Guengerich, F. P.; Arneson, K. O.; Williams, K. M.; Deng, Z.; Harris, T. M. *Chem. Res. Toxicol.* **2002**, *15*, 780-792.
8. Guengerich, F. P.; Voehler, M.; Williams, K. M.; Deng, Z.; Harris, T. M. *Chem. Res. Toxicol.* **2002**, *15*, 793-798.
9. Sabbioni, G.; Skipper, P. L.; Büchi, G.; Tannenbaum, S. R. *Carcinogenesis* **1987**, *8*, 819-824.
10. Sabbioni, G. *Chem. Biol. Interact.* **1990**, *75*, 1-15.
11. Ueng, Y.-F.; Kuwabara, T.; Chun, Y.-J.; Guengerich, F. P. *Biochemistry* **1997**, *36*, 370-381.
12. Johnson, W. W.; Guengerich, F. P. *Proc. Natl. Acad. Sci. USA* **1997**, *94*, 6121-6125.
13. Johnson, W. W.; Ueng, Y.-F.; Mannervik, B.; Widersten, M.; Hayes, J. D.; Sherratt, P. J.; Ketterer, B.; Guengerich, F. P. *Biochemistry* **1997**, *36*, 3056-3060.
14. Johnson, W. W.; Ueng, Y.-F.; Yamazaki, H.; Shimada, T.; Guengerich, F. P. *Chem. Res. Toxicol.* **1997**, *10*, 672-676.
15. Langouët, S.; Coles, B.; Morel, F.; Becquemont, L.; Beaune, P. H.; Guengerich, F. P.; Ketterer, B.; Guillouzo, A. *Cancer Research* **1995**, *55*, 5574-5579.

Chapter 13

The Use of Mammalian Cell Lines to Investigate the Role of Aldo-Keto Reductases in the Detoxication of Aldehydes and Ketones

Rachel Gardner, Shubana Kazi, and Elizabeth Ellis

Departments of Bioscience and Pharmaceutical Sciences, University of Strathclyde, Glasgow G1 1XW, Scotland, United Kingdom

Aldehydes and ketones are present in wide range of compounds including drugs, food and environmental pollutants. To study their effects on biological systems, a range of appropriate cell-based assays can be used, which investigate the nature of damage caused. The role of aldo-keto reductases in protecting against these toxic effects has been assessed in a variety of experimental systems, including cell lines that naturally express high levels of aldo-keto reductase enzymes, and constructed cell lines that express individual components of a particular detoxication pathway.

Exposure to Aldehydes and Ketones

Aldehydes and ketones are present in a diverse range of natural and synthetic compounds to which living organisms may be exposed (*1*). Many endogenous substances which are intermediates of metabolism possess aldehyde or ketone groups, including metabolites of steroid hormones, aldehydes of biogenic amines and succinic semialdehyde, a metabolite of γ-aminobutyrate (*2*). Organisms also encounter aldehydes and ketones from external sources, including commonly-consumed compounds present in the diet of higher eukaryotes, for example 2-hexenal in vegetables (*3*), methylglyoxal found in coffee (*4*), diacetyl in butter and wine (*5*). Certain environmental pollutants and their metabolites also possess aldehyde and ketone groups: examples include acrolein and crotonaldehyde, present in tobacco smoke and petrol and diesel exhaust (*6,7*); the pesticide chlordecone (*8*); chloroacetaldehyde, a breakdown product of vinyl chloride (*9*) and *trans, trans*-muconaldehyde, a metabolite of benzene (*10*). Some drugs also possess carbonyl groups: such as metyropone, adriamycin and warfarin (*11*), as do some of their metabolites, such as aldophosphamide, a metabolite of cyclophosphamide (*12*). Changes in cell physiology brought about by irradiation and oxidative stress can also lead to the production of carbonyl-containing compounds, for example the peroxidation of lipids giving rise to products such as malondialdehyde and 4-hydroxynonenal (*13*).

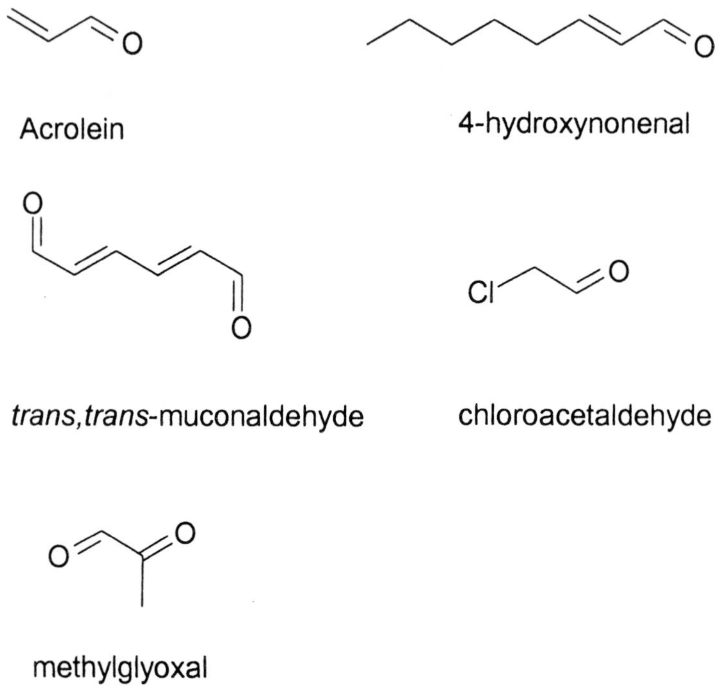

Acrolein

4-hydroxynonenal

trans,trans-muconaldehyde

chloroacetaldehyde

methylglyoxal

Figure 1. Some toxic aldehydes

Toxicity of Aldehydes and Ketones

Many aldehydes and ketones are chemically-reactive because of the electrophilicity of the C1 carbon. This makes these compounds likely to interact with nucleophilic centres in biomolecules, leading to toxic effects on biological systems. In α,β-unsaturated carbonyl compounds, the C3 position is also electrophilic, allowing further nucleophilic attack and 1,4 addition, which after rearrangement gives rise to 1,2 addition products (Michael addition) (Reviewed in (14)). Studies using cell lines have allowed investigations into the consequence of this reactivity in some considerable detail.

Cytotoxicity

Carbonyl-containing compounds can be cytotoxic as a result of their interaction with biological molecules. A wide variety of aldehydes and ketones have been shown to cause loss of viability in cultured cells (Table 1) (15).

Table 1. Cytoxicity of some aldehydes demonstrated in cell lines

Compound	Cells	Assay	Ref
4-HNE	CHO	Trypan blue exclusion	(16)
t,t-MUC	rat hepatocytes	Trypan blue exclusion	(17)
Base propenals	HeLa	Colony-forming assay	(18)
Acrolein	rat hepatocytes	Trypan blue exclusion	(19)
ALLN	CHO	Colony-forming assay	(20)
Methylglyoxal	PC12	Methylene blue staining	(21)
Benzaldehyde	Hepa1c17	Colony-forming assay	(15)
Chloroacetaldehyde	rat hepatocytes	Trypan Blue exclusion	(9)
Acrolein	V79	MTT assays	(22)

4-HNE, 4-hydroxynonenal; t,t-MUC, trans,trans-muconaldehyde; ALLN, N-acetyl-leucyl-leucyl-norleucinal.

We have examined the cytotoxicity of trans-2-nonenal to a range of cell lines (Figure 2). It is apparent that there is considerable variation in different cells response to aldehydes and ketones. For example, the rat liver line BL8 appears to have much greater intrinsic resistance to this compound than does the hamster lung fibroblast line V79.

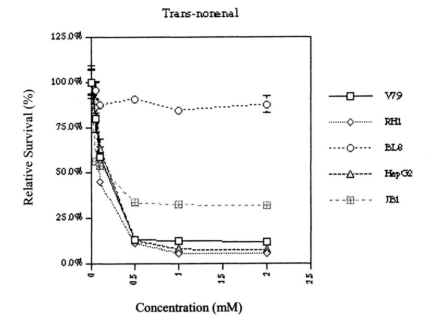

Figure 2. Cytotoxicity of trans-2-nonenal to a variety of cell lines. Cells were cultured in the presence of increasing doses of aldehydes and survival measured relative to untreated cells. JB1, BL8 and RH1 are rat liver cell lines (23,24), V79 is a hamster lung cell line (25) and HepG2 is a human liver cell line (26).

The use of cytoxicity assays gives an indication of the level of cellular damage caused by a particular aldehyde, but does not reveal the nature of the damage, nor give any information on potential pathways of detoxication. Some studies have been able to show the presence of aldehyde-protein adducts within the cell, which can also be quantitated and used as an indicator of the type of cell damage (Table 2).

Table 2. Detection of aldehyde-protein adducts in cell lines

Aldehyde	Antibodies	Cells	Ref
4-HNE	raised to 4-HNE-conjugate	rat hepatocytes	(27)
Acrolein	raised to acrolein-conjugates	mouse hepatocytes	(28)
Crotonaldehyde	raised to CA conjugates	mouse hepatocytes	(29)

4-HNE, 4-hydroxynonenal; CA, crotonaldehyde.

Treatment with some toxic aldehydes has also been shown to lead to a depletion in glutathione levels because of their reactivity with this important thiol. Compounds include chloroacetaldehyde (9), formaldehyde (30) and 4-hydroxynonenal (31). Consequential to glutathione depletion is the production of oxygen radicals, leading to oxidative stress which is thought to occur after treatment of cells with trans,trans-muconaldehyde (32).

Despite these various observable parameters of cellular damage, relatively little is known of the mechanisms of cell death caused by aldehydes. However, acrolein and methylglyoxal have been shown to trigger apoptosis-like events in cell lines (33,34). 4-Hydroxynonenal, which has also been shown to cause apoptosis (35), has been shown to rapidly activate c-Jun N-terminal protein kinase (JNK) as well as SEK1, an upstream kinase of JNK in PC-12 cells (36). Similarly, methylglyoxal has been show to activate JNK in Jurkat T cells (37). The consequences of activation of these pathways on cell proliferation and survival are far-reaching, and suggest that even sub-lethal exposure to certain aldehydes can lead to perturbations of cellular function.

Genotoxicity

In addition to cytotoxic effects, some aldehydes are known to be mutagenic through their interaction with DNA, causing DNA-aldehyde adducts and/or mutations (8-11,13,38) (Table 3) and this has been examined in several ways in cultured cells (39-41). Several aldehydes are known to form adducts with DNA in cell-based systems, which gives a clear and measurable indication of their mutagenic potential (Table 4).

Table 3. Mutagenicity of Aldehydes Determined Using Cell Lines

Compound	Cells	Assay	Type of Damage	Ref
Formaldehyde	A549	PFGE	Double-strand breaks	(42)
Methylglyoxal	COS-7	plasmids	Tranversions	(43)
Methylglyoxal	T-lymphocytes	hprt		(44)
Chloroacetaldehyde	B-lymphocytes	hprt	Deletions	(45)
t,t-MUC	V79	hprt	Point	(46)
Acrolein	Fibroblasts	plasmids	Substitutions	(47)
n-alkanals	V79	hprt	Point	(48)
2-alkenals	V79	Comet	Double-strand breaks	(41)

Abbreviations: t,t-MUC, trans,trans-muconaldehyde; PFGE, pulsed field gel electrophoresis; HPRT, hypoxanthine-guanine phosphoribosyl transferase.

Table 4. DNA-aldehyde Adducts Detected in Cell Lines

Compound	Cells	Assay	Ref
Chloroacetaldehyde	B-lymphoid	HPLC-fluorimetry	(49)
Acetaldehyde	epithelial cells	^{32}P-postlabelling	(50)
methylglyoxal	HL60	^{14}C-labelling	(51)
methylglyoxal	epithelial cells	^{32}P-postlabelling	(52)
acrolein	COS-7	shuttle plasmids	(53)

Pathways of Aldehyde and Ketone Metabolism

In order to counter the potentially lethal effects of exposure to aldehydes, organisms have evolved several enzyme systems to detoxify reactive aldehydes and ketones (Figure 3). Well-established pathways include oxidation by

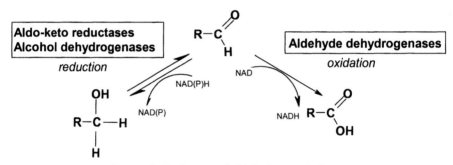

Figure 3. Pathways of aldehyde metabolism

aldehyde dehydrogenases to acids (1), and for α,β-unsaturated carbonyls, conjugation to glutathione, carried out by glutathione S-transferases (54). A third pathway of detoxication is the reduction of these compounds to alcohols which, through further detoxication steps, can be excreted from the organism (55). Several enzyme families are known to be involved in the reduction of aldehydes and ketones, including the alcohol dehydrogenases, which constitute the short-chain dehydrogenase/reductases (SDR) (56) and the medium-chain dehydrogenase/ reductases (57). Over recent years, a major family of enzymes carrying out the reduction of aldehydes and ketones known as aldo-keto reductases (AKR) has been described (58). The roles of many of the AKR enzymes in detoxication of aldehydes and ketones are now being elucidated through a combination of molecular biological and cell biological approaches. Many studies have examined the levels and activity of the individual enzymes present in the cell, as these will determine the ultimate fate of the compound. In some cases it has been clearly demonstrated that particular AKRs are involved in the activation of carcinogens, rather than their detoxication (63).

Identification of Aldo-Keto Reductases in Cell Lines

Carbonyl-reducing activity against a variety of substrates has been described in a variety of mammalian cell lines (59). Although useful in assessing the capacity of a particular cell line to metabolize a specific aldehyde, this overall measurement gives little indication of the roles of individual enzymes. In some cases, the activity observed is attributable to short-chain dehydrogenase/ reductases rather than aldo-keto reductases. We have measured the ability of different cell lines to reduce aldehydes and ketones at fixed substrate concentrations (Table 5). This data shows that there is considerable variability, not only in the levels of AKR activity that can be detected between the different cell lines, but also in the substrate-specificity of this activity. For example, levels of activity towards the model AKR substrate 4-NBA varies between cell lines, differences that may reflect differences either in the original source of the cells or in aberrant regulation these lines.

Table 5. Carbonyl-reducing activities in Cell Lines

Substrate	HepG2 Human (26)	V79 Hamster (25)	RH1 Rat liver (23)	JB1 Rat liver (24)	BL8 Rat liver
4-NBA	0.39	0.87	0.71	1.31	0.50
2-CBA	0.07	0.10	0.06	0.25	0.12
SSA	3.80	29.1	20.2	21.1	4.15
D-glucose	0.12	0.24	0.09	0.11	0.20
DL-glyceraldehyde	0.38	0.17	1.20	0.31	0.16
9,10PQ	5.76	32.5	34.7	39.0	16.7
MG	4.46	12.1	34.0	95.0	12.1

Activities are expressed as nmole/min/mg of total cell protein and were measured using substrate concentrations at 1mM with NADPH as cofactor. 4-NBA, 4-nitrobenzaldehyde; 2-CBA, 2-carboxybenzaldehyde; SSA, succinic semialdehyde; 9,10PQ, 9,10-phenanthrenequinone; MG, methylglyoxal.

Often the spectrum of enzymes and activities observed in tumor-derived cell lines differs significantly from normal tissue, which may account for the observation that some tumor cells are more resistant to the cytotoxic effects of aldehydes. In some cases, increased resistance can be directly correlated with this increased enzyme activity (60). Cell lines that are resistant to particular aldehydes and show increased AKR activity have enabled the purification of enzymes whose overexpression leads to resistance. In this way, an aldo-keto reductase was purified from a Chinese hamster ovary derived cell line that was resistant to a cytotoxic tripeptidyl aldehyde (ALLN). The purification of individual enzymes followed by peptide sequencing has allowed their identification. For example, an aldo-keto reductase that was elevated in ethacrynic acid resistant human colon cells was purified and identified as a dihydrodiol dehydrogenase (AKR1C1). Its presence at high levels in the cells increased the ability to reduce ethacrynic acid and thus prevent its toxicity (61).

More recently, the availability of AKR-specific antibodies or cDNAs has meant that it has been possible to detect not only which AKR are present, but also the level of each AKR in the cell. For example, antibodies raised to rat aldose reductase AKR1B1 showed it to be present in cell lines derived from hepatomas, whereas it is not normally present in hepatocytes (62). Similarly, Western blots have shown that AKR1C proteins are expressed at high levels in human lung carcinoma (A549) cells, and the use of isoform-specific reverse transcriptase-PCR showed that AKR1C1, AKR1C2, and AKR1C3 were all expressed in this line. In this case, the expression of these enzymes is associated with an increased capacity to activate polycyclic aromatic hydrocarbons to carcinogens rather than detoxication of aldehydes (63).

Enzymes of the AKR7 family of aldo-keto reductases are known to be capable of detoxifying aflatoxin B_1 dialdehyde as well as other toxic aldehydes and ketones (64,65). We have examined the expression of AKR7-related proteins in cell lines in an attempt to link the activities for particular substrates with the presence of these different enzyme activities. The blot in Figure 4 reveals the presence of AKR7-related proteins in three rat liver cell lines (RH1, JB1 and BL8)(23,24), HepG2 cells (26) and V79 hamster lung cells (25). Comparison of the levels of AKR7-related proteins in Figure 4 and AKR7-specific activities in Table 5 shows that two of the cell lines that express high levels of AKR7 (JB1 and BL8) also have high levels of reductase activity towards 2-carboxybenzaldehyde, a relatively-specific AKR7 substrate (66). On the other hand, V79 cells that express lower levels of AKR7-related proteins show lower levels of 2-CBA reductase activity. However, RH1 cells and HepG2 cells also have relatively high levels of AKR7, yet lower levels of 2-CBA reductase activity. This can be explained by the presence of non-AKR7-related 2-CBA reductase activity in the JB1 and BL8 cell lines that augments the AKR7-dependent activity observed in RH1 and HepG2 cells.

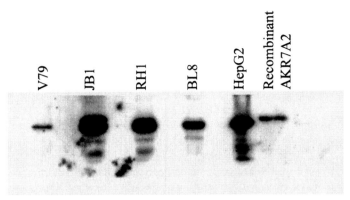

Figure 4. Western blots showing endogenous expression of AKR7-related proteins in different cell lines. Total cell extracts were separated on SDS-PAGE, blotted to membranes and probed with antibodies raised to human AKR7A2.

Construction of Stable Cell Lines Expressing Individual AKR

Because of the inherent problems in determining the role of individual AKR in detoxication, some investigators have used inhibitors to eliminate the activity of certain enzymes in cell lines. For example, the effect of an aldose reductase inhibitor SNK-860 was used to assess the role of aldose reductase in glucose-induced apoptosis in in cultured bovine retinal pericytes (*67*). This approach is applicable only to those enzymes for which specific inhibitors exist. However, the assessment of an individual enzyme's role in drug metabolism or detoxication has been achieved through the availability of cloned cDNAs encoding the enzymes. This has allowed the genes encoding these enzymes to be used in heterologous systems, thereby allowing cell lines to be constructed that express individual AKR.

One of the first studies to use this approach involved expressing the gene encoding rat aldehyde reductase AKR1A3 in rat pheochromocytoma PC12 cells, which normally express low levels of the enzyme (*36*). Cells expressing aldehyde reductase were more resistant to the cytotoxic aldehydes methylglyoxal and 3-deoxyglucosone than control cells, suggesting that this enzyme is capable of protecting neural cells against their cytotoxicity. These compounds are elevated during hyperglycemia and the results obtained present good evidence that the enzyme is involved in detoxication in a physiological context (*21*).

More recently we have constructed a cell line that stably-expresses a cDNA encoding an aldo-keto reductase of the AKR7 family in hamster lung V79 cells

(*68*). These cells normally express low levels of AKR7-related protein (Figure 3). Cells expressing rat liver AKR7A1 at levels that are similar to those found in ethoxyquin treated rat liver showed increased resistance to acrolein, as measured by reduced cytotoxicity using MTT assays and colony-forming assays (*68*). In addition, these cells were also significantly resistant to acrolein-induced mutagenicity (*68*). This result is surprising, considering that acrolein is not a particularly good substrate for AKR7A1 (*69*) but demonstrates that the ability of an enzyme to contribute to detoxication depends not only on the activity of the enzyme for a particular substrate but also the levels of the enzyme within the cell coupled with the effective concentration of substrate within the cell.

AKR are not just involved in detoxication reactions, and this has been clearly demonstrated with rat 3α-hydroxysteroid dehydrogenase/dihydrodiol dehydrogenase (AKR1C9). Expression of this enzyme in human breast MCF-7 cells led to the activation of polycyclic aromatic hydrocarbons, and increased cell death (*70*). This is because this enzyme can convert (±)-*trans*-7,8-dihydroxy-7,8-dihydrobenzo[*a*]pyrene to the cytotoxic benzo[*a*]pyrene-7,8-dione. It highlights the importance of developing a range of cell lines that allow the roles of individual enzymes in detoxication of specific compounds to be investigated.

Future Developments

The aim of using cell lines in toxicology is to develop appropriate and representative models for examining molecular aspects of toxicity that reduce the need for whole animal studies. Future work will not only examine the extent to which individual aldo-keto reductases contribute to protection against aldehyde toxicity, but also determine the levels of expression required to eliminate sub-lethal damage in these cells, using measurements of cell toxicity and mutagenicity for particular compounds, as well as more subtle indicators of cellular damage, such as activation of signalling pathways. This is particular important in the case of the inducible enzymes AKR1C1 and AKR7A1, because their role as potential chemopreventive enzymes depends on the degree of their induction in particular tissues (*65,71*). In the absence of effective induction regimes in tissue culture lines, in many cases because the relevant transcriptional regulatory machinery is absent, the use of cell lines that overexpress these enzymes can mirror the effects of elevated levels that are observed after treatment of animals with inducers, and provides a means of investigating chemoprevention at the molecular level.

In addition, it is envisaged that the construction of competitive pathways, involving several different classes of aldehyde-metabolizing enzymes will be possible. For example, the coexpression of both aldo-keto reductase and aldehyde dehydrogenase enzymes (*72*) in a single cell will allow the fate of particular aldehyde to be examined. This will allow better predictive models of metabolism to be made as to whether a particular compound will be oxidized or reduced in a given cell system, and will also allow manipulation of extraneous factors such as redox state, glutathione levels and cofactor requirement.

Similarly, for carcinogens that are known to require an aldehyde intermediate, it will be possible to co-express cytochrome P450s in the same cell, thus allowing compounds to be activated *in situ*, and providing better models of metabolism. Such pathway construction has been successfully employed in the case of compounds that require intracellular activation followed by sulfation (*73*) or conjugation by glutathione S-transferases (*74*). It is hoped that this approach will allow the roles of aldo-keto reductases in detoxication to be fully understood, and will lead to better predictive models of toxicity.

References

1. Sladek, N. E.; Manthey, C. L.; Maki, P. A.; Chang, Z.; Landkamer, G. L. *Drug Metab. Rev.* **1989**, *20*, 697-720.
2. Maser, E. *Biochem. Pharmacol.* **1995**, *49*, 421-440.
3. Eder, E.; Schuler, D. *Arch. Toxicol.* **2000**, *74*, 642-648.
4. Kasai, H.; Kumeno, K.; Yamaizumi, Z.; Nishimura, S.; Nagao, M.; Fujita, Y.; Sugimura, T.; Nukaya, H.; Kosuge, T. *Gann.* **1982**, *73*, 681-683.
5. Fornachon JC; B., L. *J. Sci. Food. Agric.* **1965**, *16*, 710-716.
6. Izard, C.; Libermann, C. *Mutat. Res.* **1978**, *47*, 115-138.
7. Lipari, F.; Sivarin, S. J. *J. Chromatogr.***1982**, *297*, 297-306.
8. Molowa, D.; Wrighton, S.; Blanke, R.; Guzelian, P. *J. Toxicol. Environ. Health* **1986**, *17*, 375-384.
9. Sood, C.; OBrien, P. *Biochem. Pharmacol.* **1994**, *48*, 1025-1032.
10. Chang, R.; Wong, C.; Kline, S.; Conney, A.; Goldstein, B.; Witz, G. *Environ. Molec. Mutag.* **1994**, *24*, 112-115.
11. Bachur, N. *Science* **1976**, *193*, 595-597.
12. Parekh, H. K.; Sladek, N. E. *Biochem. Pharmacol.* **1993**, *46*, 1043-1052.
13. Esterbauer, H.; Cheeseman, K.; Dianzani, M. P., G; Slater, T. *Biochem. J.* **1982**, *208*, 129-140.
14. Witz, G. *Free Rad. Biol. Med.* **1989**, *7*, 333-349.
15. Bassi, A. M.; Penco, S.; Canuto, R. A.; Muzio, G.; Ferro, M. *Drug Chem. Toxicol.* **1997**, *20*, 173-187.
16. Brambilla, G.; Sciaba, L.; Faggin, P.; Maura, A.; Marinari, U. M.; Ferro, M.; Esterbauer, H. *Mutat. Res.* **1986**, *171*, 169-176.
17. Goon, D.; Matsuura, J.; Ross, D. *Chem-Biol. Interact.* **1993**, *88*, 37-53.
18. Grollman, A. P.; Takeshita, M.; Pillai, K. M. R.; Johnson, F. *Cancer Res.* **1985**, *45*, 1127-1131.

19. Silva, J. M.; O'Brien, P. J. *Arch. Biochem. Biophys.* **1989**, *275*, 551-558.
20. Inoue, S.; Sharma, R. C.; Schimke, R. T.; Simoni, R., D *J. Biol. Chem.* **1993**, *268*, 5894-5898.
21. Suzuki, K.; Koh, Y. H.; Mizuno, H.; Hamaoka, R.; Taniguchi, N. *J. Biochem.* **1998**, *123*, 353-357.
22. Kazi, S.; Ellis, E. M. *Chem Biol Interact* **2002**, *140*, 121-135.
23. Kaulek, V.; Saas, P.; Alexandre, E.; Grant, H.; Richert, L.; Jaeck, D.; Tiberghien, P.; Wolf, P.; Azimzadeh, A. *Cell. Transplant.* **2001**, *10*, 739-747.
24. Manson, M. M.; Green, J. A.; Neal, G. E. *Int. J. Cancer* **1984**, *34*, 869-874.
25. Chu, E. H.; Brimer, P.; Jacobson, K. B.; Merriam, E. V. *Genetics* **1969**, *62*, 359-377.
26. Knowles, B. B.; Howe, C. C.; Aden, D. P. *Science* **1980**, *209*, 497-499.
27. Uchida, K.; Szweda, L. I.; Chae, H.-Z.; Stadtman, E. R. *Proc. Natl. Acad. Sci. USA* **1993**, *90*, 8742-8746.
28. Burcham, P. C.; Fontaine, F. *J Biochem. Mol. Toxicol.* **2001**, *15*, 309-316.
29. Fontaine, F. R.; Dunlop, R. A.; Petersen, D. R.; Burcham, P. C. *Chem. Res. Toxicol.* **2002**, *15*, 1051-1058.
30. Teng, S.; Beard, K.; Pourahmad, J.; Moridani, M.; Easson, E.; Poon, R.; O'Brien, P. J. *Chem-Biol. Interact.* **2001**, *130-132*, 285-296.
31. Hartley, D. P.; Ruth, J. A.; Petersen, D. R. *Arch. Biochem. Biophys.* **1995**, *316*, 197-205.
32. Shen, H. M.; Ong, C. N.; Shi, C. Y. *Toxicology* **1995**, *99*, 115-23.
33. Nardini, M.; Finkelstein, E. I.; Reddy, S.; Valacchi, G.; Traber, M.; Cross, C. E.; van der Vliet, A. *Toxicology* **2002**, *170*, 173-185.
34. Sheader, E. A.; Benson, R. S.; Best, L. *Biochem. Pharmacol.* **2001**, *61*, 1381-1386.
35. Haynes, R. L.; Szweda, L.; Pickin, K.; Welker, M. E.; Townsend, A. J. *Mol. Pharmacol.* **2000**, *58*, 788-794.
36. Song, B. J.; Soh, Y.; Bae, M.; Pie, J.; Wan, J.; Jeong, K. *Chem. Biol. Interact.* **2001**, *130-132*, 943-954.
37. Du, J.; Suzuki, H.; Nagase, F.; Akhand, A. A.; Yokoyama, T.; Miyata, T.; Kurokawa, K.; Nakashima, I. *J. Cell. Biochem.* **2000**, *77*, 333-344.
38. Marnett, L. J.; Hurd, H. K.; Hollstein, M. C.; Levin, D. E.; Esterbauer, H.; Ames, B. N. *Mutat. Res.* **1985**, *148*, 25-34.
39. Bradley, M. O. **1981**.
40. Singh, N. P.; McCoy, M. T.; Tice, R. R.; Schneider, E. L. *Exp. Cell. Res.* **1988**, *175*, 184-191.
41. Glaab, V.; Collins, A. R.; Eisenbrand, G.; Janzowski, C. *Mutat. Res.* **2001**, *497*, 185-197.
42. Vock, E. H.; Lutz, W. K.; Ilinskaya, O.; Vamvakas, S. *Mutat. Res.* **1999**, *441*, 85-93.

43. Murata-Kamiya, N.; Kamiya, H.; Kaji, H.; Kasai, H. *Mutat. Res.* **2000**, *468*, 173-182.
44. Hou, S. M.; Nori, P.; Fang, J. L.; Vaca, C. E. *Environ. Mol. Mutagen.* **1995**, *26*, 286-291.
45. Chiang, S. Y.; Swenberg, J. A.; Weisman, W. H.; Skopek, T. R. *Carcinogenesis* **1997**, *18*, 31-36.
46. Chang, R. L.; Wong, C. Q.; Kline, S. A.; Conney, A. H.; Goldstein, B. D.; Witz, G. *Environ. Mol. Mutagen.* **1994**, *24*, 112-115.
47. Curren, R. D.; Yang, L. L.; Conklin, P. M.; Grafstrom, R. C.; Harris, C. C. *Mutat. Res.* **1988**, *209*, 17-22.
48. Brambilla, G.; Cajelli, E.; Canonero, R.; Martelli, A.; Marinari, U. M. *Mutagenesis* **1989**, *4*, 277-279.
49. Zielinski, B.; Hergenhahn, M. *Fresenius J Anal Chem* **2001**, *370*, 97-100.
50. Vaca, C. E.; Nilsson, J. A.; Fang, J. L.; Grafstrom, R. C. *Chem. Biol. Interact.* **1998**, *108*, 197-208.
51. Kang, Y.; Edwards, L. G.; Thornalley, P. J. *Leuk. Res.* **1996**, *20*, 397-405.
52. Vaca, C. E.; Fang, J. L.; Conradi, M.; Hou, S. M. *Carcinogenesis* **1994**, *15*, 1887-1894.
53. Kanuri, M.; Minko, I. G.; Nechev, L. V.; Harris, T. M.; Harris, C. M.; Lloyd, R. S. *J. Biol. Chem.* **2002**, *277*, 18257-18265.
54. Hayes, J. D.; Pulford, D. J. *Crit. Rev. Biochem. Mol. Biol.* **1995**, *30*, 445-600.
55. Wermuth, B. In *Enzymology of Carbonyl Metabolism*; Flynn, T. G., Weiner, H., Eds.; Alan R Liss: New York, 1985; Vol. 2.
56. Jornvall, H.; Persson, B.; Krook, M.; Atrian, S.; Gonzalez-Duarte, R.; Jeffery, J.; Ghosh, D. *Biochemistry* **1995**, *34*, 6003-6013.
57. Nordling, E.; Jornvall, H.; Persson, B. *Eur. J. Biochem.* **2002**, *269*, 4267-4276.
58. Jez, J. M.; Flynn, T. G.; Penning, T. M. *Biochem. Pharmacol.* **1997**, *54*, 639-647.
59. Gebel, T.; Maser, E. *Biochem. Pharmacol.* **1992**, *44*, 2005-2012.
60. Takahashi, M.; Fujii, J.; Miyoshi, E.; Hoshi, A.; Taniguchi, N. *Intl. J. Cancer* **1995**, *62*, 749-754.
61. Ciaccio, P. J.; Stuart, J. E.; Tew, K. D. *Mol Pharmacol* **1993**, *43*, 845-853.
62. Takahashi, M.; Hoshi, A.; Fujii, J.; Miyoshi, E.; Kasahara, T.; Suzuki, K.; Aozasa, K.; Taniguchi, N. *Japanese J. Cancer Res.h* **1996**, *87*, 337-341.
63. Palackal, N.; Lee, S.; Harvey, R.; Blair, I.; Penning, T. *J Biol Chem* **2002**, *277*, 24799-24808.
64. Judah, D. J.; Hayes, J. D.; Yang, J. C.; Lian, L. Y.; Roberts, G. C. K.; Farmer, P. B.; Lamb, J. H.; Neal, G. E. *Biochem. J.* **1993**, *292*, 13-18.
65. Ellis, E. M.; Judah, D. J.; Neal, G. E.; Hayes, J. D. *Proc. Natl. Acad. Sci. USA* **1993**, *90*, 10350-10354.

66. Ireland, L. S.; Harrison, D. J.; Neal, G. E.; Hayes, J. D. *Biochem. J.* **1998**, *332*, 21-34.
67. Naruse, K.; Nakamura, J.; Hamada, Y.; Nakayama, M.; Chaya, S.; Komori, T.; Kato, K.; Kasuya, Y.; Miwa, K.; Hotta, N. *Exp. Eye. Res.* **2000**, *71*, 309-315.
68. Gardner, R. Li, D Ellis, E.M.. Manuscript in preparation
69. Ellis, E. M.; Hayes, J. D. *Biochem. J.* **1995**, *312*, 535-541.
70. Tsuruda, L.; Hou, Y.; Penning, T. *Chem. Res. Toxicol.* **2001**, *14*, 856-862.
71. Ciaccio, P. J.; Jaiswal, A. K.; Tew, K. D. *J. Biol. Chem.* **1994**, *269*, 15558-15562.
72. Townsend, A. J.; Leone-Kabler, S.; Haynes, R. L.; Wu, Y.; Szweda, L.; Bunting, K. D. *Chem-Biol. Interact.* **2001**, *130-132*, 261-273.
73. Glatt, H.; Bartsch, I.; Christoph, S.; Coughtrie, M. W.; Falany, C. N.; Hagen, M.; Landsiedel, R.; Pabel, U.; Phillips, D. H.; Seidel, A.; Yamazoe, Y. *Chem. Biol. Interact.* **1998**, *109*, 195-219.
74. Fields, W. R.; Morrow, C. S.; Doehmer, J.; Townsend, A. J. *Carcinogenesis* **1999**, *20*, 1121-1125.

Aldo-Keto Reductases, the Stress Response, and Cell Signaling

Chapter 14

Aldose Reductase and the Stress Response

Aruni Bhatnagar[1], Si-Qi Liu[1], Sanjay Srivastava[1], Kota V. Ramana[2], and Satish K. Srivastava[2]

[1]Division of Cardiology, Department of Medicine, University of Louisville, Louisville, KY 40202
[2]Department of Human Biological Chemistry and Genetics, University of Texas Medical Branch, Galveston, TX 77555

Aldose reductase (AKR1B1, abbreviated as AR) is a member of the aldo-keto reductase (AKR) superfamily. It catalyzes the reduction of a wide range of aldehydes including aldo sugars. While the physiological role of AR is unclear, its broad substrate selectivity suggests that the enzyme may be involved in the detoxification of both endogenous and environmental aldehydes. The detoxification role of AR is supported by the observations that the enzyme is readily induced by electrophilic and oxidative stress, as well as by disease conditions associated with increased generation of reactive oxygen species (ROS). Tissue abundance of AR is also increased upon stimulation by growth factors and cytokines that increase ROS as a part of their post-receptor signaling. The mechanisms by which stress affects AR have not been identified, although both transcriptional and post-translational mechanisms have been proposed to mediate an increase in the enzyme protein and activity, and to alter its subcellular localization. The adaptive significance of AR relates to the observations that inhibition of this enzyme exacerbates oxidative injury in cellular and *in vivo* models of oxidative stress. In culture systems, inhibition of AR sensitizes cells to hydrogen peroxide, aldehydes derived from lipid peroxidation and glucose metabolism, and drugs generating reactive carbonyls. The antioxidant role of AR has also been demonstrated in *in vivo* models of myocardial ischemia and vascular injury and inflammation. Collectively, this evidence is consistent with the view that AR plays a critical antioxidant role in preventing electrophilic injury caused by toxic aldehydes.

Aldose reductase (AKR1B1, abbreviated as AR) is a broad specificity aldo-keto reductase (AKR) that catalyzes the reduction of a wide range of substrates containing an aldehyde functional group (*1,2*). Although the enzyme was initially discovered as a glucose reducing agent in seminal vesicles (hence, the name *aldose* reductase), extensive substrate-specificity studies have shown that AR is capable of catalyzing the reduction of several structurally unrelated aldehydes (*1*). Indeed, glucose is a rather poor substrate of the enzyme and given the wide array of aldehydes that are substrates of this enzyme, it appears that under most conditions, reduction of glucose is not the primary or even a significant role of AR. Nonetheless, *in vivo*, the enzyme does seem to participate in glucose reduction, since inhibition of AR prevents intracellular sorbitol accumulation (*1,2*). It is currently believed that under euglycemic conditions, AR does not play a particularly significant role in glucose metabolism, however, during hyperglycemia, AR could channel excessive glucose through the polyol pathway, thereby preventing the overloading of other glucose-metabolizing pathways. However, continued glucose reduction by AR leads to tissue injury such as that associated with the development of secondary diabetic complications (*1,2* and also see Srivastava *et al.*, this volume).

Although its incidental involvement in diabetic complications has received extensive attention, the high level of AR expression in tissues such as skeletal muscle or heart, which do not face intracellular hyperglycemia even during diabetes, suggests that the enzyme may be involved in other physiological processes. These functions of AR have not been clearly delineated, but must relate to the substrate specificity and its ability to catalyze efficiently the reduction of unique aldehydes. However, multiple exhaustive searches have failed to identify a single unique substrate-product pair that is specifically reduced by AR (*3,4*). Indeed, it has been suggested that there *is* no specific substrate of AR and that the structure of the enzyme has evolved to bind and catalyze a broad array of aldehydes of variable structure; making it a non-specific aldehyde reductase (*5*).

Specificity is one of the characteristic features of enzyme catalysis. Most enzymes recognize only a few substrates, because specific interactions between an enzyme and its substrate generate the energy required to overcome the thermodynamic barriers to catalysis. Aldose reductase, on the other hand, represents an altogether different paradigm of enzyme catalysis. Most of the energy required for AR-mediated catalysis is derived from tight binding to

NADPH (5). Because no energetic demands are placed on substrate binding, cofactor-bound AR could catalyze the reduction of any aldehyde that could fit into the active site, and could interact with the active site residues in an appropriate conformation without the energy-requiring movement of side chains. As a result of this adaptation, AR could catalyze the reduction of small aromatic and aliphatic aldehydes such as para-nitrobenzaldehyde and methylglyoxal (3), as well as bulky aldehydes such as isocorticosteroids (6) and aldehyde-glutathione conjugates (4). Nonetheless, some minimal requirements and preferred motifs that determine AR-mediated catalysis have been identified. These include, for small aldehydes, the presence of an oxidized carbon or a hydroxyl group at C-2 position, and for bulky aldehydes, the potential for multiple hydrophobic interactions between the active site side chains of the enzyme and the substrate. The requirements for ionic interactions are minimal, but could dramatically affect AR catalysis. This is best illustrated by the relative efficiencies with which the enzyme catalyzes the reduction of C-3 aldehydes. The C-3 aldehydes that have an OH (e.g. glyceraldehyde) or phenyl (substituted or unsubstituted benzaldehydes) groups or those that contain an oxidized C-2 carbon (e.g., methyl glyoxal) are reduced much more efficiently than those that contain a simple alkyl chain (e.g. propanal or acrolein) (3). The requirement for ionic interactions at the C-2 position is overcome by the high hydrophobicity of bulky aliphatic aldehydes and, therefore, the enzyme readily reduces compounds such as C-6 to C-12 alkanals, alkenals, and 4-hydroxyalkenals with comparable catalytic efficiencies (4). Catalysis of even bulkier molecules such as steroids requires additional interactions that are provided by the relatively open active site and substrate selecting interactions of the C-terminus domain. Although the mechanism for steroid binding to the protein has not been probed, this must be related to the interactions of steroids with other AKR proteins such as the 3α/20α hydroxysteroid dehydrogenase (7). The binding and catalysis of glutathione conjugates represents yet another aspect of the plasticity of the AR active site, and it involves interactions with ionic groups removed from the active site but at the lip of the β-barrel, providing an anchor that selects a specific binding orientation of glutathione but does not participate or interfere with the binding of smaller unsubstituted aldehydes (see, Srivastava *et al.*, this volume).

While AR has been shown to catalyze the reduction of several aldehydes, the physiological and pathological significance of this enzyme to electrophile detoxification has not been fully assessed. For instance, AR has been shown to catalyze the reduction of isocorticosteroids (6) and progesterone (10), which implicates its involvement in steroid metabolism. This role is supported by the high abundance of AR in steroidogenic tissues such as adrenal glands. However, whether AR contributes to steroid bioactivity or metabolism remains to be assessed. Similarly, AR has been implicated in the metabolism of biogenic amines. *In vitro* the enzyme catalyzes the reduction of 3,4-

dihydrophenylacetaldehyde (DOPAL), which is toxic metabolite formed by the oxidative deamination of dopamine (*11*). DOPAL could be either oxidized by aldehyde dehydrogenase or reduced by AR, and combined inhibition of both these enzymes has been recently reported to increase rotenone toxicity, which is associated with DOPAL accumulation (*12*). In addition, AR has also been shown to catalyze the reduction of a range of aldehydes generated during lipid peroxidation (*13-15*), and in this case there is extensive evidence (*vide infra*) supporting the view that the enzyme may be involved in detoxifying and removing toxic lipid-derived aldehydes and thus protecting against one major component of oxidative injury.

The surprising aspect of AR catalysis is not that it binds to several substrates, but that it catalyzes their reduction so well. Not only does the enzyme make little energetic demands from aldehyde binding, but it also interacts productively with several aldehydes of remarkably diverse structure. This feature of AR catalysis is precisely what would be expected of a detoxification enzyme that could be recruited to participate in multiple detoxification pathways. Indeed, a growing body of evidence demonstrates that AR is upregulated by stress and that it succeeds in diminishing the toxicity of several endogenous and environmental aldehydes.

Stress-Induced Upregulation of AR

Several types of stress responses have been shown to result in AR upregulation. The classical example of this phenomenon is osmotic stress, which results in a profound (20 to 30-fold) increase in the transcription of the AR gene (*17,18*). This aspect of AR is related to its ability to catalyze the reduction of glucose to sorbitol. Upregulation of AR in high salt medium results in the accumulation of sorbitol, which, being a compatible osmolyte, balances the intracellular and extracellular osmotic gap, and is of physiological significance in maintaining the tonicity of the renal inner medullary cells exposed to high salt. The molecular events associated with the increase in AR by osmotic stress are relatively well understood and involve the activation of TonEBP or NFAT-5 (*19*). Upon activation, this transcription factor is translocated to the nucleus (*20*) where it stimulates the transcription of several osmo-protective genes, including AR. The osmoregulatory role of AR appears to be a phylogentically well conserved response, and even in plants an increase in AR or related proteins (e.g., dehydrin) is associated with increased salt-resistance (*21*). Recently, a stress-activated AR has been cloned from alfalfa, and ectopic expression of this gene in tobacco has been shown to enhance tolerance to oxidative stress and dehydration (*22*). The simultaneous protection against osmotic and oxidative stress, conferred by AR upregulation, suggests that this enzyme may be part of the underlying mechanism for the phenomenon of cross-tolerance, whereby adaptation to one type of stress (i.e., high salt or dehydration) imparts resistance to other forms of stress (i.e., oxidative injury). This is illustrated by Figure 1

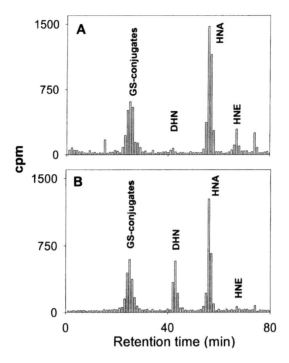

Figure 1. Osmotic stress stimulates HNE metabolism. Confluent rat aortic smooth muscle cells were cultured in normal DMEM (A) or DMEM containing 300 mM NaCl (B) for 24 h and were then incubated with 8 μM [³H]-HNE for 30 min at 37 °C. After the incubation period, the medium was removed and the radioactivity was separated by HPLC using a reverse-phase C₁₈ column. On the basis of the retention time of synthetic metabolites, 4 major peaks in the eluate were identified to be: glutathionyl conjugates of HNE (GS-conjugates), 1,4-dihydroxynonene (DHN), 4-hydroxynonanoic acid (HNA) and unmetabolized HNE. The chemical identity of these peaks was established by electrospray or gas chromatography mass spectrometry as described before (15,16). Note that treatment with high salt, which stimulates AR expression (17,18), increases the formation of the AR-derived product DHN. Also, a greater decrease in the HNE peak in the high salt treated cells (compared to those cultured in iso-osmotic medium) suggests that upregulation of AR increases HNE consumption and metabolism without significantly affecting the formation of other metabolites (i.e., the glutathione conjugates and HNA).

which shows that culturing vascular smooth muscle cells in high salt increases AR expression and facilitates 4-hydroxy-*trans*-2-nonenal (HNE) metabolism.

Although AR-dependent cross-tolerance between osmotic and oxidative stress in animals remains to be demonstrated, several studies show an increase in AR upon exposure to oxidative stress. For instance, increased expression of an AR-like protein is a significant component of the responses of yeast to hydrogen peroxide (*23*). Although the specific function of this gene product has not been examined, based on the evidence collected from mammalian tissues, it appears to be part of an antioxidant response for protecting yeast from the harmful effects of aldehydes generated by peroxide-induced oxidation of cell membranes. Interestingly, the yeast AR encoded by the GRE3 gene is upregulated by high salt, hydrogen peroxide, heat-stress, and carbon starvation (*24*). Overexpression of GRE3 increases methylglyoxal tolerance and complements the deficiency of the glyoxalase system, indicating that the yeast AR may be involved in methylglyoxal metabolism (For further discussion on yeast genes see Chang *et al.*, this volume). Whether AR plays a similar role in mammalian systems is unclear, but induction of the AR gene by methylglyoxal in vascular smooth muscle cells has been demonstrated, and inhibition of AR has been shown to increase methylglyoxal toxicity (*25*, see also Van der Jagt *et al.*, this volume).

Consistent with its role as a stress-induced antioxidant enzyme, AR has been shown to be upregulated under conditions associated with increased generation of reactive oxygen species such as diabetes (*26,27*), vascular inflammation (*28*) and injury (*29*), heart failure (*30*), and myocardial ischemia (*31*) and liver cirrhosis (*32*). Though the normal liver shows little or no expression of AR, liver samples from patients with alcoholic liver disease show increased AR immunoreactivity. The mechanisms of this induction are not known but may relate to increased cytokine generation during alcoholic liver disease because stimulation of hepatocytes with TNF-α increases AR protein and mRNA (*33*). The increase in hepatic AR could also be related to generalized oxidative stress, since exposure of HepG2 cells to either iron overload, hydrogen peroxide, or HNE increases AR mRNA (*34*). Thus, changes in cytokines and oxidative stress could contribute to the upregulation of AR in diseased liver.

The abundance of AR in the heart, which constitutively expresses high levels of AR, is also increased by stress. In tissue samples from human failing hearts, AR was one of the few genes whose expression was consistently increased (*30*), suggesting that AR could be part of the cardioprotective response to myocardial stress. However, in animal models of heart failure (*35*), we observe a decrease in AR mRNA, protein, and activity. These changes were accompanied by a parallel increase in the accumulation of lipid peroxidation-derived aldehydes, supporting the view that diminished AR activity could be related to the increased oxidative stress observed in the failing heart. A parallel decrease in TonEBP activity in these hearts indicates that part of the mechanism for the decrease in AR activity may be volume overload and the decreased ionic strength in the interstitial spaces of the failing heart. The link between volume changes and AR expression could also account for the differences observed with human and rodent hearts. While one reason could be the treatment of humans with heart failure using multiple drugs that could increase AR, the other could be that humans with failing hearts are usually treated to compensate for changes in volume, in which case AR expression is increased in response to oxidative stress. Contrastingly, in animal models, there is a profound volume overload, and so the abundance of AR is decreased due to a decrease in extracellular tonicity. This view, if validated, would suggest that preventing volume overload could diminish oxidative stress (by preserving or stimulating AR), and that injuries due to osmotic and oxidative stress are interlinked.

An increase in the AR activity has also been demonstrated in hearts subjected to ischemia *ex vivo* (*36*) or *in situ* (*31*). Interestingly, both transcriptional and post-translational mechanisms seem to contribute to the upregulation of AR in the ischemic heart. Increased AR activity was measured in rat hearts subjected to low flow ischemia for 50 min. However, no change in AR protein was observed, suggesting that this increase in AR activity was due to post-translational modification of AR by a nitric oxide-dependent mechanism. Recent results also show that in rabbit hearts, brief episodes of ischemia lead to an increase in AR activity 24 h later and that this increase in AR was prevented by inhibiting NO synthase or protein kinase C. Significantly, the increase in AR was restricted to the membrane fraction, indicating that ischemia promotes the translocation of AR from the cytoplasm to the membrane. A similar increase in the membrane-associated AR activity was also observed acutely in rat hearts subjected to global ischemia and reperfusion *ex vivo* (Figure 2), suggesting that translocation of AR to the membrane may be an acute response to stress, and may be essential for removing toxic lipid-aldehydes localized to the membrane. While such post-translational changes in AR have not been extensively examined in other tissues, differences in the kinetic properties of AR isolated from diabetic and non-diabetic tissues have been reported (*37,38*, and also see Srivastava, *et al.*, this volume). Together, changes in the transcription of the AR gene, post-translational modification of the protein and changes in its subcellular localization underscore the importance of this enzyme in protecting the heart from ischemic injury.

AR-Mediated Protection Against Oxidative Injury

In contrast to multiple studies showing changes in AR gene or protein in response to stress, there is a marked paucity of studies probing the protective role of AR against oxidative injury. *In vivo* data are available only for cardiovascular injury, although protective effects of AR against hydrogen

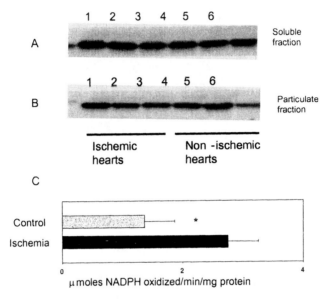

Figure 2. Increased membrane association of AR in ischemic hearts. Isolated rat hearts were perfused ex vivo in the Langendorff mode and were subjected to global ischemia for 30 min and reperfusion for 60 min. Control hearts were perfused aerobically for 90 min. At the end of the experiment, the tissue was homogenized and separated into particulate and soluble fractions and as described before (31). The abundance of AR was quantified by Western blot analysis using anti-AR antibodies in cytosolic or soluble fraction (A) and the particulate fraction (B). Lanes1-3 show AR in the ischemic heart and lanes 4-6 show the data from non-ischemic hearts. Panel C shows AR activity in the particulate fractions of the ischemic and non-ischemic (control) hearts (C). Note: ischemia results in modest increase in AR associated with the particulate, but not the soluble fraction of the extract and a significant increase in the particulate AR activity was observed in the ischemic hearts as compared to non-ischemic hearts, indicating that ischemia acutely upregulates AR. * P < 0.05 (n = 3).

peroxide (*39*), HNE (*40*), and methylglyoxal (*25*) have been demonstrated in cell culture systems. In muscular blood vessels of adult healthy animals, the expression of AR is restricted to the endothelium (*28,29*). During injury or inflammation however, AR is expressed at high levels in the proliferative neointima (*29*). The association of AR with proliferating cells suggests a growth regulatory role of the enzyme. Moreover, the significance of the increase in AR activity in vascular lesions is suggested by the observation that treatment with AR inhibitors decreases restenosis in balloon-injured rat carotid arteries (*29*) and diminishes intimal thickening in galactose-fed dogs (*41*). These observations are consistent with the view that in proliferative lesions, intimal upregulation of AR is essential for growth and survival of smooth muscle cells. Given the extensive evidence demonstrating the association between oxidative stress and abnormal smooth muscle cells growth, it appears that AR may be essential for removing toxic products generated during oxidative signaling.

A protective, antioxidant role of AR has also been demonstrated in the conscious rabbit model of regional myocardial ischemia *in situ* (*31*). In this model, preconditioning with ischemia upregulates AR (*vide supra*). More importantly however, inhibition of AR abrogates the protective (i.e., infarct sparing) effects of preconditioning. Interestingly, in naïve, non-preconditioned hearts, inhibition of AR did not increase infarct size, suggesting that prior stimulation, and perhaps membrane translocation of AR, is a prerequisite for the protective effects of the enzyme to manifest. Measurements of HNE reveal that inhibition of AR does not promote HNE accumulation, in the non-preconditioned heart, but does cause a marked increase in HNE in the preconditioned heart. Based on these observations we suggest AR is an obligatory mediator of preconditioning, and that it needs to be expressed in the right amount in the right location to be protective. However, the protective role of AR in the rabbit heart is in contrast to reports suggesting that inhibition of AR actually increases ischemic injury in *ex vivo* preparations of rat hearts (*44,45*). It is likely that these differences may be related to differences in the nature of *ex vivo* and *in situ* models of ischemia, and the difficulties associated with measuring ischemic injury. However, the demonstration that inhibition of AR abolishes ischemic injury *in situ*, and that this is accompanied by increased accumulation of HNE is consistent with the view that the myocardial activity of the enzyme is required to diminish the electrophilic overload generated by ischemia and reperfusion. The expression of AR was also increased in the myocardium of rat fed 6% ethanol for 6 weeks (Figure 3). Because epidemiological data suggest that consumption of low levels of ethanol protects against cardiovascular disease (*46*) and in animal models exposure to mild chronic exposure to ethanol protects the heart from ischemia-reperfusion injury (*47*), the upregulation of AR in the hearts of ethanol-fed animals suggests that, analogous to preconditioning, the cardioprotective effects of ethanol feeding may be related also to the upregulation of AR. However, additional functional studies are required to delineate the contribution of AR to the cardioprotective phenotype of ethanol-fed animals.

In summary, current evidence suggests that AR is a broad specificity aldehyde reductase that is increased in response to osmotic and oxidative stress. The enzyme seems to be particularly critical to the metabolism and detoxification of aldehydes derived from lipid peroxidation, although it could be recruited for removing other toxic aldehydes such as methylglyoxal and DOPAL, and for metabolizing glucose. Despite this evidence the constitutive physiological role of this enzyme in non-renal tissues remains unclear. The

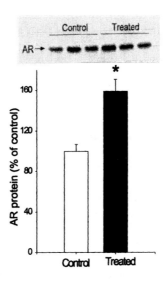

Figure 3. Ethanol feeding upregulates cardiac AR. Adult Spargue-Dawley rats were fed 6% ethanol in drinking water for 6 weeks, after which the animals were euthanized and their hearts were removed and homogenized. The abundance of AR in the myocardial homogenates was quantified by Western blot analysis. Homogenates of hearts of rats maintained without ethanol were used as control. * P < 0.05 (n = 3).

relationship between AR and stress adaptation could be further strengthened by elucidating the mechanism that regulate the expression of the AR gene during different forms of stress, and by assessing the significance of these changes to tissue injury.

Acknowledgements

This work was supported by NIH grants HL55477, HL59378, and AA13299.

References

1. Bhatnagar, A.; Srivastava, S. K. *Biochem. Med. Metabol. Biol.* **1992**, *48*, 91-121.
2. Yabe-Nishimura, C. *Pharmacol. Rev.* **1998**, *50*, 21-33.
3. Vander Jagt, D. L.; Kolb, N. S.; Vander Jagt, T. J.; Chino, J.; Marinez, F. J., Hunsaker, L. A.; Royer, R. E. *Biochem. Biophys. Acta* **1995**, *1249*, 117-126.
4. Srivastava, S.; Watowich, S. J.; Petrash, J. M.; Srivastava, S. K.; Bhatnagar, A. *Biochemistry* **1999**, *38*, 42-54.
5. Grimshaw, C. E. *Biochemistry* **1992**, *31*, 10139-10145.
6. Wermuth, B.; Monder, C. *Eur. J. Biochem.* **1983**, *131*, 423-426.
7. Ma, H.; Penning, T. M. *Proc. Natl. Acad. Sci. U.S.A.* **1999**, *96*, 11161-11166.
8. Dixit, B. L.; Balendiran, G. K.; Watowich, S. J.; Srivastava, S.; Ramana, K. V.; Petrash, J. M.; Bhatnagar, A.; Srivastava, S. K. *J. Biol. Chem.* **2000**, *275*, 21587-21595.
9. Ramana K. V.; Dixit, B. L.; Srivastava, S.; Balendiran, G. K.; Srivastava, S. K.; Bhatnagar, A. *Biochemistry* **2000**, *39*, 12172-12180.
10. Warren, J. C.; Murdock, G. L.; Ma, Y.; Goodman, S. R.; Zimmer, W. E. *Biochemistry* **1993**, *32*, 1401-1406.
11. Kawamura, M.; Eisenhofer, G.; Kopin, I. J.; Kador, P. F.; Lee, Y. S.; Tsai, J. Y.; Fujisawa, S.; Lizak, M. J.; Sinz, A.; Sato S. *Biochem. Pharmacol.* **1999**, *58*, 517-524.
12. Lamensdorf, I.; Eisenhofer, G.; Harvey-White, J.; Nechustan, A.; Kirk, K.; Kopin I. J. *Brain Res.* **2000**, *868*, 191-201.
13. Vander Jagt, D. L.; Kolb, N. S.; Vander Jagt, T. J.; Chino, J.; Martinez, F. J.; Hunsaker, L. A.; Royer, R. E. *Biochim. Biophys. Acta.* **1995**, *1249*, 117-126.
14. Srivastava, S.; Chandra A.; Bhatnagar, A.; Srivastava, S. K.; Ansari, N. H. *Biochem. Biophys. Res. Comm.* **1995**, *217*, 741-746.
15. Srivastava, S.; Chandra, A.; Wang, L.-F.; Seifert, W. E.; DaGue, B. B.; Ansari, N. H.; Srivastava, S. K., Bhatnagar, A. *J. Biol. Chem.* **1998**, *273*, 10893-10900.
16. Srivastava, S.; Conklin, D. J.; Boor, P. J.; Liu, S. Q.; Srivastava, S. K.; Bhatnagar, A. *Atherosclerosis* **2001**, *18*, 339-350.
17. Burg, M. B.; Kwon, E. D.; Kultz, D. *Ann. Rev. Physiol.* **1997**, *59*, 437-455.
18. Burg, M. B.; Kwon, E.D.; Kultz, D. *FASEB J.* **1996**, *10*, 1598-1606.
19. Miyakawa, H.; Woo, S. K.; Dahl, S. C.; Handler, J. S.; Kwon, H. M. *Proc. Natl. Acad. Sci. U.S.A.* **1999**, *96*, 2538-2542.
20. Dahl, S. C.; Handler, J. S.; Kwon, H. M. *Am. J. Physiol.* **2001**, *280*, C248-C253.

21. Maestri, E.; Malcevschi, A.; Massari, A.; Marmiroli, N. *Mol. Gen. Genom. MGG.* **2002**, *267*, 186-201.
22. Bartels D. *Trends Plant Sci.* **2001**, *6*, 284-286.
23. Godon, C.; Lagniel, G.; Lee, J.; Buhler, J. M.; Kieffer, S.; Perrot, M.; Boucherie, H.; Toledano, M. B.; Labarre J. The H_2O_2 stimulon in Saccharomyces cerevisiae. *J. Biol. Chem.* **1998**, *273*, 22480-22489.
24. Aguilera, J.; Prieto, J. A. *Curr. Genet.* **2001**, *39*, 273-283.
25. Chang, K. C.; Paek, K. S.; Kim, H. J.; Lee, Y. S.; Yabe-Nishimura, C.; Seo, H. G. *Mol. Pharmacol.* **2002**, *61*, 1184-1191.
26. Ghahary, A.; Chakrabarti, S.; Sima, A. A.; Murphy L. J. *Diabetes.* **1991**, *40*, 1391-1396.
27. Hodgkinson, A. D.; Sondergaard, K. L.; Yang, B.; Cross, D. F.; Millward, B. A.; Demaine, A. G. *Kidney Internatl.* **2001**, *60*, 211-218.
28. Rittner, H. L.; Hafner, V.; Klimiuk, P. A.; Szweda, L. I.; Goronzy, J. J.; Weyand, C. M. *J. Clin. Invest.* **1999**, *103*, 1007-1013.
29. Ruef, J.; Liu, S.-Q.; Bode, C.; Tocchi, M.; Srivastava, S.; Runge, M.; Bhatnagar, A. *Arterio. Thromb. Vasc. Biol.* **2000**, *20*, 1745-1752.
30. Yang, J.; Moravec, C. S.; Sussman, M. A.; DiPaola, N. R.; Fu, D.; Hawthorn, L.; Mitchell, C. A.; Young, J. B.; Francis G. S.; McCarthy, P. M.; Bond, M. *Circulation.* **2000**, *102*, 3046-3052.
31. Shinmura, K.; Bolli, R.; Liu, S.-Q.; Tang, X. L.; Kodani, E.; Xuan, Y.-T.; Srivastava, S. K.; Bhatnagar, A. *Circ. Res.* **2002**, *91*, 240-246.
32. O'Connor, T.; Ireland, L. S.; Harrison, D. J.; Hayes, J. D. *Biochem. J.* **1999**, *343*, 487-504.
33. Iwata, T.; Sato, S.; Jimenez, J.; McGowan, M.; Moroni, M.; Dey, A.; Ibaraki, N.; Reddy, V.N.; Carper, D. *J. Biol. Chem.* **1999**, *274*, 7993-8001.
34. Barisani, D.; Meneveri, R.; Ginelli, E.; Cassani, C.; Conte, D. *FEBS Lett.* **2000**, *469*, 208-212.
35. Srivastava, S.; Chandrasekar, B.; Bhatnagar, A.; Prabhu, S.D. *Am. J. Physiol.* (In Press) **2002**.
36. Hwang, Y. C.; Sato, S.; Tsai, J. Y.; Yan, S.; Bakr, S.; Zhang, H.; Oates, P.J.; Ramasamy, R. *FASEB J.* **2002**, *16*, 243-245.
37. Srivastava, S. K.; Hair; G. A.; Das, B. *Proc. Natl. Acad. Sci. U.S.A.* **1985**, *82*, 7222-7226.
38. Chandra, D.; Jackson, E. B.; Ramana, K. V.; Srivastava, S. K.; Bhatnagar, A. *Diabetes* (In Press) **2002**.
39. Spycher, S. E.; Tabataba-Vakili, S.; O'Donnell, V. B.; Palomba, L.; Azzi, A. *FASEB J.* **1997**, *11*, 181-188.
40. Spycher, S.; Tabataba-Vakili, S.; O'Donnell, V. B.; Palomba, L.; Azzi, A. *Biochem. Biophys. Res. Comm.* **1996**, *226*, 512-516.

41. Kasuya, Y.; Ito, M.; Nakamura, J.; Hamada, Y.; Nakayama, M.; Chaya, S.; Komori, T.; Naruse, K.; Nakashima, E.; Kato, K.; Koh, N.; Hotta N. *Diabetologia,* **1999**, *42, 1404-1409.*
42. Ushio-Fukai, M.; Alexander, R. W.; Akers, M.; Yin, Q.; Fujio, Y.; Walsh, K.; Griendling, K. K. *J. Biol. Chem.* **1999**, *274,* 22699-22704.
43. Irani K. *Circ. Res.* **2000** *87,* 179-183.
44. Trueblood, N.; Ramasamy, R. *Am. J. Physiol.* **1998**, *275,* H75-H83.
45. Ramasamy, R.; Liu, H.; Oates, P. J.; Schaefer, S. *Cardiovas. Res.* **1999**, *42,* 130-139.
46. Muntwyler, J.; Hennekens, C. H.; Buring, J.E.; Gaziano, J. M. Lancet **1998**, *352,*1882-1885.
47. Miyamae, M.; Diamond, I.; Weiner, M. W.; Camacho, S. A.; Figueredo, V. M. *Proc. Natl. Acad. Sci. U.S.A.* **1997**, *94,* 3235-3239.

Chapter 15

Aldose Reductase Regulates Reactive Oxygen Species Mediated–Inflammatory Signals

Kota V. Ramana[1], Deepak Chandra[1], Sanjay Srivastava[2],
Aruni Bhatnagar[2], Bharat B. Aggarwal[3], and Satish K. Srivastava[1]

[1]Department of Human Biological Chemistry and Genetics, University of Texas Medical Branch, Galveston, TX 77555
[2]Division of Cardiology, Department of Medicine, University of Louisville, Louisville, KY 40202
[3]Cytokine Research Laboratory, Department of Bioimmunotherapy, M. D. Anderson Cancer Center, University of Texas, Houston, TX 77030

Aldose reductase (AKR1B1, abbreviated as AR) is an aldo-keto reductase (AKR) that catalyzes the reduction of glucose to sorbitol, which is the first and rate-limiting step of the polyol pathway. The enzyme is also efficient in reducing multiple lipid peroxidation-derived aldehydes and their glutathione conjugates. However, the physiological significance of AR catalysis in intracellular signaling and cell cycle regulation has not been assessed. Our recent observations suggest that AR mediates the mitogenic and cytotoxic signals of reactive oxygen species generated by growth factors as well as cytokines. The mitogenic role of AR is supported by the observations that inhibition of AR attenuates TNF-α, PDGF or bFGF, but not phorbol ester (PMA)-mediated activation of the redox-sensitive transcription factors – NF-κB and AP1. *In vivo*, these stimuli are key regulators of vascular smooth muscle cell proliferation and apoptosis of vascular endothelial cells. Pretreatment with AR inhibitors also prevented the activation of protein kinase C (PKC) induced by TNF-α, bFGF, and PDGF but not PMA, indicating that inhibition of the PKC/NF-κB pathway may be a significant cause of the anti-inflammatory effects of AR inhibitors. Collectively, these observations suggest that AR may be a critical and essential mediator of vascular inflammation and that it may be required to facilitate multiple signaling events regulating cell cycle progression and apoptosis.

Oxidative and Antioxidant Roles of Aldose Reductase

Accumulating evidence demonstrating that the inhibition of aldose reductase (AKR1B1, abbreviated as AR) prevents or delays hyperglycemic injury in experimental models of diabetes, supports the view that AR is etiologically involved in the development of secondary diabetic complications such as cataractogenesis, retinopathy, neuropathy, nephropathy and microangiopathy (1-6). Based on this evidence, it has been proposed that the increased flux of glucose via AR causes osmotic and oxidative changes, which in turn, trigger a sequence of metabolic changes resulting in gross tissue dysfunction, altered intracellular signaling, and extensive cell death (1,2). Thus, during hyperglycemia, AR could induce oxidative stress due to a decrease in the NADPH/NADP$^+$ ratio caused by the reduction of excessive glucose to sorbitol (5,6). This oxidative stress could be intensified further by subsequent metabolism of sorbitol to fructose via sorbitol dehydrogenase, which converts NADH to NAD$^+$, thereby inducing a state of chemical hypoxia.

Our previous work shows that, in addition to glucose, AR also catalyzes the reduction of lipid peroxidation-derived aldehydes and their glutathione (GSH) conjugates (7-9). The purified enzyme displays high affinity for these aldehydes and their conjugates. In isolated perfused hearts, erythrocytes, cultured vascular smooth muscle cells (VSMC), and vascular endothelial cells (VEC), inhibition of AR prevents the reduction of glutathiolated aldehydes (10-12). Because these aldehydes, particularly the α,β-unsaturated alkenals such as 4-hydroxy *trans* 2-nonenal (HNE) are highly reactive and cytotoxic (13), their metabolism via AR may be an antioxidant defense mechanism, at least under euglycemic conditions. The antioxidant role of AR is consistent with the observations that in a variety of cell types, AR is upregulated by several oxidants such as hydrogen peroxide (14), methyl glyoxal (15), and nitric oxide (16), and during pathological conditions associated with oxidative stress, such as myocardial ischemia and reperfusion (17). The expression of AR is also enhanced upon exposure to its substrates such as HNE (14) suggesting that conditions associated with high accumulation of electrophiles lead to induction of the AR gene.

Several lines of experimental evidence suggest that up-regulation of AR by oxidative stress may be a protective response. The expression of AR in VSMC, for example, is increased by H_2O_2 or HNE stimulation and inhibition of the enzyme increases the sensitivity of these cells to both H_2O_2 and HNE -mediated cell death *(14)*. An *in vivo* correlation to these observations is provided by Rittner et al., who demonstrate that during giant cell arteritis, elevated levels of AR are associated with the proliferative intima and with areas of high positive immunoreactivity with anti-protein-HNE antibodies *(18)*. Interestingly, AR inhibitors increased HNE accumulation in this model and caused a 4-fold elevation of apoptotic cells in the vascular lesions. These data suggest that increased AR activity may be specifically associated with areas of high inflammation or growth and that the activity of this enzyme may be essential for cell cycle progression. Because in vascular tissues, redox signaling is a significant component of these events (inflammation, cell growth and survival), we examined whether changes in the AR activity affect proinflammatory responses or apoptosis.

Aldose Reductase Mediates the Mitogenic and the Cytotoxic Signals

A mitogenic role of AR is suggested by our observation that treatment with AR inhibitors diminishes the extent of neointimal hyperplasia in the balloon-injured rat carotid arteries *(19)*. Furthermore, the involvement of AR in mediating cell growth and survival is supported by the results of our cell culture studies showing that inhibition of this enzyme by structurally distinct inhibitors prevents the proliferation of VSMC and the apoptosis of VEC in response to TNF-α and growth factors. High levels of growth factors and cytokines are expressed in the intimal lesions *in vivo* and these mediators have been shown to play a critical role in the progression of atherosclerosis and restenosis *(20-22)*. In restenotic vessels, TNF-α is the most abundant cytokine and intimal proliferation of VSMC has been linked to increased TNF-α stimulation. The mitogenic role of TNF-α in vascular lesions is in contrast to its well-known pro-apoptotic activity seen with several cell types in culture. Hence, because of its central role in mediating vascular inflammation and abnormal growth, we examined the effects of TNF-α on VSMC proliferation in culture and tested whether this paradigm of VSMC growth depends on AR. When serum-starved rat aortic VSMC were stimulated with TNF-α in culture a two-fold increase in cell growth was observed within 24 h (Figure 1). This is consistent with previous reports documenting the proliferative activity of TNF-α with VSMC in culture *(23,24)*. However, when the serum-starved cells were pre-incubated with AR inhibitors, sorbinil or tolrestat, 12 to 24 h before TNF-α stimulation, a significant prevention of TNFα-induced proliferation was observed. Neither of the AR inhibitors affected cell growth by itself, i.e., in the absence of TNF-α stimulation. These data clearly demonstrate that inhibition of AR prevents TNF-α-induced VSMC growth in culture (Figure 1).

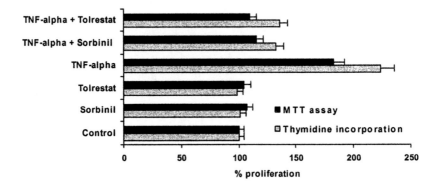

*Figure 1. Aldose reductase inhibitors attenuate TNF-α–induced VSMC proliferation *P<0.5 versus TNF-α alone.*

The antimitogenic effects of AR inhibition suggest that the enzyme may be involved in promoting growth signaling and in facilitating cell cycle progression. An alternate explanation could be that inhibition of AR can induce cell death and that the observed inhibition of VSMC proliferation induced by TNF-α in the presence of AR inhibitors may be due to progress towards apoptosis. The apoptotic consequences of abnormal cell cycle progression have been demonstrated with several cell types. Indeed, treatment with AR inhibitors has been shown to be associated with an increase in apoptosis *(18)*. Therefore, we examined whether pre-incubation with AR inhibitors diminishes cell viability and induces cell death. We measured cell viability by nucleosomal degradation and the induction of apoptosis by the activation of caspase-3. Our results show that treatment with AR inhibitors does not cause a significant increase in cell death as assessed by both the nucleosomal degradation and caspase-3 assay (data not shown). Thus, collectively our results suggest that inhibition of AR prevents TNF-α-induced VSMC cell proliferation and that this effect is associated specifically with a decrease in AR activity and is not due to the non-specificity of a single inhibitor.

It has been reported that inhibition of AR prevents hyperproliferation and hypertrophy of VSMC cultured in the presence of high glucose *(25)*. No mechanism was proposed for the antimitogenic effects of AR inhibitors on hyperglycemic cell growth. Our results suggest that the AR-dependence of VSMC growth may be more general than previously anticipated and that inhibition of the enzyme prevents growth even in normoglycemic media. Mechanisms by which inhibition of AR prevents VSMC growth remain unclear, however, one mechanism is that AR may regulate the redox state of the cell. Redox changes are central to cell growth and the induction of apoptosis, and changes in the generation and removal of reactive oxygen species (ROS) are key events associated with mitogenic as well as apoptotic signaling *(26,27)*. Although the specific molecular events associated with ROS signaling have not been identified, antioxidant interventions have been shown to prevent VSMC growth and interrupt apoptotic signaling in several cell types *(26-28)*. Moreover, because of the high propensity of membrane lipids to undergo oxidation, and high bioactivity of the products of lipid peroxidation, changes in membrane lipid composition could couple ROS generation to intracellular signaling.

The oxidation of membrane lipids generates a variety of active end products, of which the α,β-unsaturated aldehydes could be particularly significant *(13)*. Unsaturated aldehydes are the major end products of lipid peroxidation, and due to their high reactivity elicit a variety of biological responses including induction of the heat-shock response, stress-activated signaling and changes in gene expression *(29,30)*. In the absence of growth factors or even ROS, lipid peroxidation generated aldehydes such as 4-hydroxy-*trans*-2-nonenal (HNE) are competent and complete VSMC mitogens *(31)*. Because AR catalyzes the reduction of these aldehydes and their glutathione conjugates, it is possible that the mitogenic role of AR catalysis is related to its ability to remove and detoxify lipid peroxidation products generated by a ROS-mediated component of growth factor signaling. In this view, AR catalysis represents a protective antioxidant mechanism that prevents the accumulation of ROS-generated toxicants. This role will be analogous to that of phosphatases in phosphorylation signaling cascades. Given the "early" generation of ROS and the delayed-early upregulation of AR, we envision the following sequence of events: stimulation by growth factors rapidly increases ROS generation which would initiate lipid peroxidation leading to the generation of lipid-derived aldehydes that sustain mitogenic signaling. However, the resultant oxidative stress upregulates AR, and as the AR levels increase, the concentrations of the lipid-derived aldehydes diminish, and the redox signaling is turned off, so that cell growth continues unimpeded.

Aldose Reductase Inhibitors Prevent the Activation of Redox-Sensitive Transcription Factors

A redox-sensitive, pro-mitogenic role of AR is supported by our observations that pretreatment with the AR inhibitors - sorbinil or tolrestat, attenuates cytokine and growth factor- induced activation of the redox-sensitive transcription factors NF-κB and AP1 in VSMC as well as VEC. Activation of the transcription factor NF-κB is a fundamental route in the ROS as well as the TNF-α-mediated signal transduction pathway (27). Therefore, we determined whether AR prevents TNF-α-mediated activation of NF-κB and AP1. As shown in Figure 2, pretreatment with AR inhibitors prevents TNF-α-induced activation of NF-κB and AP1. Inhibition of TNF-α–induced activation of NF-κB and AP1 by AR inhibitors was dose-dependent, and either of these drugs at 10 μM prevented the activation of these transcription factors. However, for maximal inhibition, a minimum of 12 h of pre-incubation with the AR inhibitors was required. Incubation with either sorbinil or tolrestat in the absence of TNF-α did not inhibit NF-κB activation, suggesting that AR inhibitors by themselves do not significantly affect NF-κB activation, but attenuate cell proliferation only when the cells are stimulated with TNF-α or other growth factors. These results suggest that AR plays a pivotal role in the signal transduction pathway of TNF-α leading to the activation of the transcription factor NF-κB.

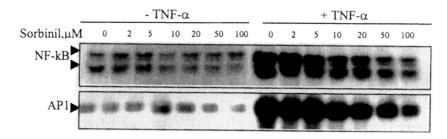

Figure 2. Aldose reductase inhibitor, sorbinil inhibits redox sensitive transcription factors NF-κB and AP1 in VSMC.

The transcription factor NF-κB regulates the expression of a large number of genes that encode critical proteins involved in inflammation such as cytokines (TNF-α, IL-1, IL-8, IL-6), cell adhesion proteins like ICAM-1, VCAM and enzymes such as AR, nitric oxide synthase, cyclo-oxygenase, and Mn-SOD (32-34). Thus, our observation that inhibition of AR prevents the NF-κB activation suggests that some of the anti-diabetic effects of AR inhibitors may be related to the suppression of pro-inflammatory factors, such as cytokines and adhesion molecules. Moreover, the ability of these inhibitors to prevent cellular events associated with NF-κB activation (i.e., cell growth) suggests a central role of AR as a critical determinant of inflammatory responses and cytokine generation.

Aldose Reductase Mediates Signals Upstream to PKC

To probe the mechanism by which inhibition of AR prevents the activation of NF-κB, we examined signaling events upstream to the activation of this transcription factor. In unstimulated cells, NF-κB is sequestered in the cytoplasm by association with Iκ-Bα (*32-34*). Upon stimulation, Iκ-Bα is phosphorylated and degraded by the proteosome, allowing the dissociated NF-κB to migrate to the nucleus to activate gene transcription. Since we observed that both Iκ-Bα-phosphorylation as well as its proteolytic degradation were blocked by inhibiting AR (data not shown), it appears that inhibition of AR does not directly affect the NF-κB system, but interrupts signaling events further upstream to the phosphorylation and degradation of Iκ-Bα and the translocation of NF-κB to the nucleus (*35*). A key event in the phosphorylation of Iκ-Bα and stimulation of the entire NF-κB pathway appears to be the activation of protein kinase C, because it has been shown that treatment of VSMC with protein kinase inhibitors decreases NF-κB activity (*36*). Therefore AR-dependence of PKC activation could account for decreased IκB-α-phosphorylation in the cells treated with sorbinil or tolrestat. Indeed, as shown in Figure 3, we found that pre-treatment with AR inhibitors prevented PKC activation.

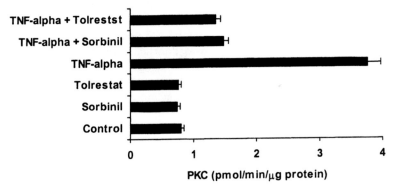

Figure 3. Aldose reductase inhibitors attenuate TNF-α–induced total membrane bound PKC activity in the VSMC.

The extent to which inhibition of AR prevents PKC activation is independent on the PKC stimulus, since comparable inhibition was observed when PKC was activated either by TNF-α, FGF, PDGF or Ang-II (data not shown), indicating that inhibition of AR could prevent PKC activation regardless of the specific receptor activated. However, AR inhibitors were unable to prevent PKC activation when the kinase was directly activated by phorbol esters. These data clearly identify the AR-sensitive locus in growth factor and cytokine mediated activation of the NF-κB pathway to be just upstream of PKC activation but after growth factor receptor activation. In addition, the lack of effect of sorbinil and tolrestat on PMA-induced activation of PKC suggests that these compounds do not cause non-specific inhibition of either PKC or its downstream signaling events (activation of NF-κB) but that their effects are restricted to metabolic/signaling pathways that require AR or its reaction products.

Several lines of evidence link PKC activation to AR activity *(4,5)*. Both PKC and AR are coordinately upregulated *(37)* and recent proteomic analysis shows that at least in the heart, PKC and AR are part of the same signaling complex *(38)*, and inhibition of AR by epalrestat attenuates the VSMC proliferation by high glucose by suppressing the PKC activity *(39)*. However, the AR-dependence of PKC activation is not entirely clear. It has been suggested that AR activity is required for the synthesis of diacylglycerol, which is an obligatory cofactor for the activation of classical and novel PKC isoforms. This view is consistent with our observation that pre-incubation with AR inhibitors was essential for inhibiting both the PKC and NF-κB activation, suggesting that the lack of AR activity imposes a metabolic deficiency that interferes with optimal activation of PKC. In addition to regulating DAG synthesis, AR could also affect PKC activity by modulating the cellular redox state. Several studies demonstrate that growth factors such as FGF, PDGF and thrombin exert their mitogenic activity by generating ROS in VSMC and in other cell types *(27)*. Since, the ROS can initiate lipid peroxidation that leads to the generation of bioactive unsaturated aldehydes such as HNE, the induction of lipid peroxidation and the generation of its products may be linked to the mitogenic effects of ROS. Thus, by regulating accumulation of lipid peroxidation AR, could indirectly affect PKC activity.

The effects of AR inhibitors on TNF-α signaling could be further accentuated by hyperglycemia. As shown in Figure 4, chronic hyperglycemia decreases $NADPH/NADP^+$ ratio due to increased reduction of glucose to sorbitol catalysed by AR. This decrease in NADPH availability could limit the activity of γ-glutamylcysteine synthase, which is a key enzyme in glutathione synthesis, resulting in a decrease in the intracellular antioxidant - glutathione. Indeed, diabetic or TNF-α-stimulated cells do show a decrease in intracellular glutathione levels, which could by itself cause activation of NF-κB by inducing oxidative stress *(40)*. This view is supported by the observations that treatment of VSMC with N-acetyl cysteine (a thiol antioxidant and a potential precursor of glutathione) attenuates NF-κB activation in cells exposed to high glucose or cytokines *(41)*.

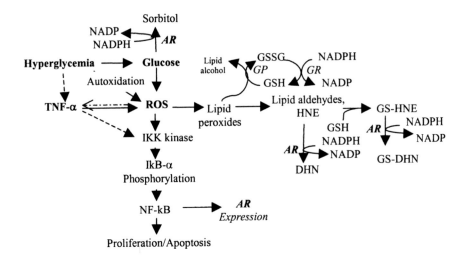

Figure 4. Possible roles of AR in TNF-α and hyperglycemia-induced signaling cascades. ROS, reactive oxygen species; GP, glutathione peroxidase; GR, glutathione reductase

Hence, increased AR activity could impair the glutathione (GSH) redox cycle and treatment with AR inhibitors could restore glutathione synthesis, thereby attenuating the effects of the TNF-α or hyperglycemia. Further, the accelerated polyol pathway during TNF-α or hyperglycemia, by perturbing pyridine coenzyme nucleotide metabolism, suppresses nitric oxide synthase and initiates the synthesis of diacylglycerol which could lead to the activation of PKC and consequently NF-κB.

Conclusions

In conclusion, our results show that treatment with AR inhibitors, sorbinil and tolrestat, diminishes TNF-α induced activation of NF-κB and the proliferation of VSMC in culture. The anti-mitogenic effect of sorbinil was comparable in magnitude to the ability of this drug to inhibit Iκ-Bα phosphorylation, indicating that inhibition of AR prevents the phosphorylation of Iκ-Bα by inhibiting the kinase responsible for Iκ-Bα phosphorylation. Furthermore, our results show that inhibition of AR prevents PKC activation and may be a significant mechanism by which AR inhibitors abrogate NF-κB signaling and diminish cell growth. Collectively, these observations are consistent with a critical role of AR in mediating or facilitating pro-inflammatory responses. Further elucidation of these mechanisms will help to delineate key intracellular signals that mediate abnormal VSMC growth during atherosclerosis and restenosis and those that contribute to the vascular complications of diabetes.

References

1. Brownlee, M. *Nature* **2001**, *414*, 813-820.
2. *Diabetic Neuropathy*; Dyck, P. J.; Thomas, P. K., Eds.; W.B. Saunders Co.; Philadelphia, PA, 1999.
3. Beckman, J. A.; Creager, M. A.; Libby, P. *JAMA* **2002**, *287*, 2570-2581.
4. Kinoshita, J. H. *Exp. Eye Res.* **1990**, *50*, 567-573.
5. Bhatnagar, A.; Srivastava, S. K. *Biochem. Med. Metabol. Biol.* **1992**, *48*, 91-121.
6. Yabe-Nishimura, C. *Pharmacol. Rev.* **1998**, *50*, 21-33.
7. Srivastava, S.; Watowich, S. J.; Petrash, J. M.; Srivastava, S. K.; Bhatnagar, A. *Biochemistry* **1999**, *38*, 42-54.
8. Dixit, B. L.; Balendiran, G. K.; Watowich, S. J., Srivastava, S., Ramana, K. V.; Petrash, J. M.; Bhatnagar, A.; Srivastava, S. K. *J. Biol. Chem.* **2000**, *275*, 21587-21595.
9. Ramana, K. V.; Dixit, B. L.; Srivastava, S.; Balendiran, G. K.; Srivastava, S. K.; Bhatnagar, A. *Biochemistry* **2000**, *39*, 12172-12180
10. Srivastava, S.; Chandra, A.; Wang, L.-F.; Seifert, W. E.; DaGue, B. B.; Ansari, N. H.; Srivastava, S. K.; Bhatnagar, A. *J. Biol. Chem.* **1998**, *273*, 10893-10900.
11. Srivastava, S.; Dixit, B. L.; Cai, J.; Sharma, S.; Hurst, H.; Bhatnagar, A.; Srivastava, S. K. *Free Radic. Biol. Med.* **2000**, *29*, 642-651.
12. Srivastava, S.; Conklin, D. J.; Boor, P. J.; Liu, S. Q.; Srivastava, S. K.; Bhatnagar, A. *Atherosclerosis* **2001**, *18*, 339-350.
13. Esterauer, H.; Schaur, R .J.; Zollner, H. *Free Radic. Biol. Med.* **1991**, *11*, 81-128.
14. Spycher, S. E.; Tabataba,-Vakili, S.; O'Donnell, V. B.; Palomba, L.; Azzi, A. *FASEB J.* **1997**, *11*, 181-188.
15. Chang, K. C.; Paek, K. S.; Kim, H. J.; Lee, Y. S.; Yabe-Nishimura, C., Seo, H. G. *Mol. Pharmacol.* **2002**, *61*, 1184-1191.
16. Seo, H. G.; Nishinaka, T.; Yabe-Nishimura, C. *Mol. Pharmacol.* **2000**, *57*, 709-717.
17. Shinmura, K.; Bolli, R.; Liu, S. Q.; Tang, X. L.; Kodani, E.; Xuan, Y. T.; Srivastava, S.; Bhatnagar, A. *Circ. Res.* **2002**, *91*), 240-246.

18. Rittner, H. L.; Hafner, V.; Klimiuk, P. A.; Szweda, L. I.; Goronzy, J. J.; Weyand, C. M. *J. Clin. Invest.* **1999**, *103*, 1007-1013.
19. Ruef, J.; Liu, S. Q.; Bode, C.; Tocchi, M.; Srivastava, S.; Runge, M. S.; Bhatnagar, A. *Arterioscler Thromb. Vasc. Biol.* **2000**, *20*, 1745-1752.
20. Luiss, A. J. Atherosclerosis. Nature **2000**, *407*, 233-241.
21. Tanaka, H.; Sukhova, G.; Schwartz, D.; Libby, P. *Arterioscler. Thromb. Vasc. Biol.* **1996**, *16*, 12-18.
22. Rectenwald, J. E.; Moldawer, L. L.; Huber, T. S.; Seeger, J. M.; Ozaki, C. K. *Circulation* **2000**, *102*, 1697-1702.
23. Selzman, C. H.; Meldrum, D. R.; Cain, B. S.; Meng, X.; Shames, B. D.; Ao, L.; Harken, A. H. *J. Surg. Res.* **1998**, *80*, 352-356.
24. Young, W.; Mahboubi, K.; Haider, A.; Li, I.; Ferreri, N. R. *Circ. Res.* **2000**, *86*, 906-914
25. Graier, W. F.; Grubenthal, I.; Dittrich, P.; Wascher, T. C.; Kostner G. M. *Eur. J. Pharmacol.* **1995**, *294*, 221-229.
26. Powis, G.; Gasdaska, J. R.; Baker, A. *Adv. Pharmacol.* **1997**, *38*, 329-359.
27. Lander, H. M. *FASEB J.* **1997**, *11*, 118-124.
28. Su, B.; Mitra, S.; Gregg, H.; Flavahan, S.; Chotani, M. A.; Clark, K. R.; Goldschmidt-Clermont, P. J.; Flavahan, N. A. *Circ. Res.* **2001**, *89*, 39-46.
29. Keller, J. N.; Mattson, M. P. *Rev. Neurosci.* **1998**, *9*, 105-116.
30. Leonarduzzi, G.; Arkan, M. C.; Basaga, H.; Chiarpotto, E.; Sevanian, A.; Poli, G. *Free Radic. Biol. Med.* **2000**, *28*, 1370-1378.
31. Reuf J.; Rao, G. N.; Li F. Z.; Bode, C.; Patterson, C.; Bhatnagar, A.; Runge, M. S. *Circulation* **1998**, *97*, 1071-1078.
32. Tak, P.; Firestein, G. S. *J. Clin. Invest.* **2001**, *107*, 7-11.
33. Collins, T.; Cybulsky, M. I. *J. Clin. Invest.* **2001**, *107*, 255-264.
34. Garg, A.; Aggarwal, B. B. Nuclear transcription factor-κB as a target for cancer drug development. *Leukemia*, **2002**, *16*, 1053-1068.
35. Ramana, K. V.; Chandra, D.; Srivastava, S.; Bhatnagar, A.; Aggarwal, B. B.; Srivastava, S. K. *J. Biol. Chem.* **2002**, *277*, 32063-32070.
36. Hattori, Y.; Hattori, S.; Sato, N.; Kasai, K. *Cardiovasc Res.* **2000**, *46*, 188-197.
37. Henry, D. N.; Busik, J. V.; Brosius, F. C.; Heilig, C. W. *Am. J. Physiol.* **1999**, *277*, F97-F104.
38. Ping, P.; Zhang, J.; Pierce, W.; Bolli, R. *Circ. Res.* **2001**, *88*, 59-62
39. Nakamura, J.; Kasuya, Y.; Hamada, Y.; Nakashima, E.; Naruse, K.; Yasuda, Y.; Kato, K.; Hotta, N. *Diabetologia* **2001**, *44*, 480-487.
40. Haddad, J. J. *Biochem. Biophys. Res. Commun.* **2002**, *296*, 847-856.
41. Viedt, C.; Hansch, G. M.; Brandes, R. P.; Kubler, W.; Kreuzer, J. *FASEB J.* **2000**, *14*, 2370-2372.

Chapter 16

Aldo-Keto Reductases in the Stress Response of the Budding Yeast *Saccharomyces cerevisiae*

Qing Chang[1], Theresa Harter[1], Terry Griest[1], B. S. N. Murthy[1], and J. Mark Petrash[1,2,*]

Departments of [1]Ophthalmology and Visual Sciences and [2]Genetics, Washington University School of Medicine, St. Louis, MO 63110

Aldose reductase has been implicated as a factor in the pathogenesis of diabetic complications. However, little is known about the physiological role of this enzyme and of related aldo-keto reductases (AKRs) in humans. We recently turned to the budding yeast *Saccharomyces cerevisiae* as a model system for a functional genomics study of AKRs. Single gene deletions involving each of six putative yeast AKR open reading frames yielded strains with no measurable phenotypic difference from the wild type strain. However, a strain created by ablating open reading frames corresponding to the three most catalytically competent yeast AKRs (*YPR1*, *GRE3*, *GCY1*) demonstrated a marked increase in heat shock sensitivity. Oligonucleotide array experiments showed that the transcriptional response of the triple AKR null strain to heat shock was dramatically altered in comparison to the wild type strain. Transcription profiling of the triple AKR null strain also revealed substantial upregulation of stress response genes under basal (unstressed) growth conditions. Taken together, these results suggesting that yeast aldo-keto reductases may play an important role in modulating the stress response.

Aldose reductase (AKR1B1) is thought to play an important role in the pathogenesis of diabetic complications. It is therefore an attractive therapeutic target for strategies to prevent blindness and nerve conduction deficits often associated with diabetes mellitus. While the enzyme presents itself as an attractive drug target in diabetic patients with chronic hyperglycemia, little is known about the physiological role of this enzyme or of related aldo-keto reductases in the prediabetic state or in euglycemia. Given that effective therapy is most likely to occur when intervention is begun at the earliest stage of diabetes (1), a balance must be struck between the benefits of enzyme inhibition in diabetic target tissues and potential complications resulting from blockade of the enzyme's beneficial functions. At the present time, virtually nothing is known about the physiological role fulfilled by aldose reductase.

In mammalian systems, a gene knock out approach is often used to establish gene function. However, this approach for the aldo-keto reductases has not provided insight into their physiological roles in target tissues. Functional compensation by redundant AKRs including AKRs 1A (aldehyde reductase), 1B7 (FR-1) and 1B8 (MVDP) most likely obscures phenotypes that would otherwise emerge following ablation of a single AKR gene. As an alternate strategy, we are examining the budding yeast *Saccharomyces cerevisiae* as a model system for a functional genomics study of AKRs. A distinct advantage of this system centers on the ability to readily ablate multiple targeted genes in a single strain. In addition to providing insights into functional redundancy, this system allows us to use a genetic approach to study possible effector pathways associated with one or more individual genes.

In previous studies, we carried out BLAST analysis to identify yeast open reading frames encoding AKRs with functional similarity to human aldose reductase, also known as AKR1B1 (2). Some of these putative AKRs were functionally validated by studies of their cognate recombinant proteins. In the present study, we report preliminary evidence showing that ablation of the yeast AKR genes most functionally similar to AKR1B1 results in a unique strain of *S. cerevisiae* that shows enhanced sensitivity to stress. Studies with oligonucleotide arrays show that the triple null strain has an altered transcription profile consistent with an attenuated stress response. These data indicate that AKR-null strains may provide new insights into signaling mechanisms involving this family of proteins.

Diversity of Yeast Aldo-Keto Reductases

Yeast open reading frames that encode proteins with structural similarity to human aldose reductase were identified using the BLAST search algorithm (www.stanford.edu/saccharomyces) to query the *Saccharomyces cerevisiae* genome sequence. As shown in Table 1, *S. cerevisiae* has six open reading frames that encode proteins with a high degree of sequence similarity (50-60%)

Table 1. Putative Aldo-Keto Reductases in *Saccharyomyces cerevisiae*

ORF	AKR Family Designation	% Identity to HAR	% Similarity to HAR	Synonyms
YHR104W	2B6	42	60	*GRE3*
YOR120W	3A1	44	61	*GCY1*
YDR368W	3A2	42	57	*YPR1*
YBR149W	3C	38	60	*ARA1*
YJR096W	5F	38	56	-
YDL124W	5G	30	50	-

NOTE: HAR, human aldose reductase, AKR1B; Synonyms are gene names recognized by the yeast genetics community, "-" represents a gene synonym has not yet been designated.

SOURCE: BLAST results were collated through the YPD™ (21)

and sequence identity (30-44%) to human aldose reductase. Four of the six identified ORFs corresponded to previously annotated genes. In the following section, we summarize new and previously published data for each of the putative AKR genes in *Saccharyomyces cerevisiae*.

YHR104W (*GRE3*)

The *GRE3* gene encodes a ~37 kDa polypeptide shown by Kuhn and coworkers to be an NADPH-dependent oxidoreductase with catalytic activity toward a wide variety of aldehydes including aldoses and substituted aromatic aldehydes such as *p*-nitrobenzaldehyde (3). Using enzyme purified from a recombinant expression system, we recently confirmed that Gre3p has many kinetic similarities to human aldose reductase (2).

GRE3 was so-named on the basis of its upregulation in concert with two other genes (Genes de Respuesta a Estres) following stress (osmotic, oxidative, heat shock) of yeast cultures (4). *GRE3* transcripts are found to be markedly upregulated at the diauxic shift as well. While Gre3p is a relatively efficient catalyst of xylose reduction (K_m~25 mM), hyperosmotically-stressed yeast cells do not show an increase in xylose reductase activity despite an apparent induction of *GRE3* transcripts. That *GRE3* transcription is positively regulated by Msn2p and Msn4p, two well known stress responsive transcription factors (5), also suggests that this gene could play a role in the stress response (4). Null strains for *GRE3* show no obvious phenotype when examined under normal conditions or under conditions of osmotic or oxidative stress (4,6).

YOR120W (*GCY1*)

GCY1 was initially identified as a Galactose-inducible Crystallin-like Yeast protein) in a genomic screen for genes induced by heme (7). Gene sequencing revealed an open reading frame (YOR120W) that encoded a protein with similarity to ρ-crystallin (AKR 1C10), a protein found in abundance in the frog lens (8). Although Bandlow and coworkers failed to observe stong evidence to substantiate a catalytic role for Gcy1p (7), we and others subsequently demonstrated that the protein is a respectable aldo-keto reductase with activity toward simple aldehydes such as DL-glyceraldehyde and p-nitrobenzaldehyde (2,9). Using X-ray crystallography, Wilson and coworkers have shown that the overall structure of Gcy1p is remarkably similar to human aldose reductase, although some structural features of the active site suggest different mechanisms might regulate the catalytic sequence of the human and yeast enzymes (10). Although *GCY1* is strongly induced by galactose, the enzyme is not able to use this aldose as a substrate. Null strains for *GCY1* show no obvious phenotype when grown under normal laboratory conditions (2).

YDR368W (*YPR1*)

In the course of a study of yeast reduction of α- and β-keto esters, Nakamura and coworkers isolated and characterized an NADPH-dependent reductase activity that showed a degree of enantioselectivity that could be exploited for synthetic chemisty applications (11). Amino acid sequencing studies established the colinearity of their purified reductase and YDR368Wp and thus Yeast Putative Reductase 1 *(YPR1)*. Using enzymes purified from recombinant systems, we (2) and others (12) demonstrated the broad substrate specificity of Ypr1p. Among all yeast aldo-keto reductases, Ypr1p was the only enzyme sensitive to inhibition by sorbinil, a clinically-relevant aldose reductase inhibitor (2). Although *YPR1* is modestly induced by stress at the transcript (13) and protein (12) levels, null strains for YPR1 show no obvious phenotype when grown under normal culture conditions (2,12).

YBR149W (*ARA1*)

YBR149W (*ARA1*) encodes a subunit of arabinose dehydrogenase, a heterodimeric enzyme that catalyzes the oxidation of D-arabinose in the pathway leading to D-erthyroascorbic acid synthesis (14). Functioning predominantly in the direction of alcohol oxidation, Ara1p is one of the few aldo-keto reductases that acts as a carbonyl producer rather than carbonyl reducer. Not only does Ara1p differ in catalytic preference from the three yeast AKRs described above, but it also appears to be much less involved as a possible anti-stress gene. Transcriptional profiling studies have shown that *ARA1* transcripts are induced to a minimal extent following osmotic or oxidative stress in comparison to those derived from *GRE3, GCY1* and *YPR1* (13).

YJR096W

A polypeptide of 282 amino acids is encoded by the YJR096W open reading frame. The product of this gene has not been described in the literature, although evidence demonstrating the gene is transcriptionally active can be deduced from a various DNA microarray studies (see Microarray Global Viewer, http://www.transcriptome.ens.fr/ymgv/). For example, data from a study conducted by Causton and coworkers show that transcripts corresponding to YJR096W are induced under various stress conditions (15).

To further investigate whether YJR096Wp might fulfill a functional role in yeast metabolism, we examined the kinetic properties of the recombinant enzyme. In previous studies, we demonstrated that YJR096Wp can catalyze the NADPH-dependent reduction of various aldehydes such as DL-glyceraldehyde and D-xylose (2). As shown in Table 2, we have now determined kinetic constants for an extensive range of potential substrates. As can be seen from these data, YJR096Wp utilizes a broad variety of aldehyde containing compounds as substrates, including aromatic and aliphatic aldehydes as well as aldoses such as xylose. The best substrate examined is the substituted aromatic hydrocarbon, p-nitrobenzaldehyde, which is reduced with a catalytic efficiency approximately 3 orders of magnitude greater than all other compounds examined in this study. Catalysis was dependent on NADPH, as no activity could be detected when NADH was used as a cofactor. Likewise, we were unable to detect alcohol oxidation when YJR096Wp was assayed in the presence of NADP or NAD. These results suggest that this enzyme functions primarily as an aldehyde reductase.

YDL124W

YDL124W is another open reading frame that encodes a putative aldo-keto reductase. While no functional significance has yet been assigned this open reading frame, DNA microarray studies clearly demonstrate that the gene is transcriptionally active. Indeed, transcripts corresponding to this ORF were found to be elevated under some stress conditions (see Microarray Global Viewer, http://www.transcriptome.ens.fr/ymgv/). Following up from a previous study (2), we evaluated the catalytic potential of recombinant YDL124Wp by examining reduction reactions with a broad range of aldehyde-containing compounds listed in Table 2. Under conditions found to support activity of GRE3p, GCY1p and YPR1p, we failed to detect significant reaction rates when assays were conducted using aldose and aliphatic aldehydes, even when the enzyme was examined at a concentration likely to significantly exceed a reasonable physiological level (1-10 μM). However, relatively good reaction

Table 2: Kinetic constants for yeast aldo-keto reductases

Enzyme source	HAR	YJR096Wp	YDL124Wp
D-glucose			
k_{cat} (min^{-1})	5.2 ± 0.24	NMA	NMA
K_m (mM)	212 ± 26.7		
k_{cat}/K_m (min^{-1}mM^{-1})	0.02 ± 0.002		
DL-Glyceraldehyde			
k_{cat} (min^{-1})	62	6.6	NMA
K_m (mM)	0.07 ± 0.002	53.7 ± 2	
k_{cat}/K_m (min^{-1}mM^{-1})	885	0.1	
Xylose			
k_{cat} (min^{-1})	74.4 ± 0.36 64	1.0	Trace
K_m (mM)	10.2 ± 1.49 16	116 ± 6	
k_{cat}/K_m (min^{-1}mM^{-1})	7320 ± 810 4	9E-3	
p-Nitrobenzaldehyde			
k_{cat} (min^{-1})	28.1 ± 0.07	88	Trace
K_m (mM)	0.016 ±0.0017	0.5 ± 0.06	
k_{cat}/K_m (min^{-1}mM^{-1})	1784 ± 161.3	174	
Benzaldehyde			
k_{cat} (min^{-1})	26.6 ± 2.64	3.9	NMA
K_m (mM)	0.073 ± .0113	47.1 ± 6	
k_{cat}/K_m (min^{-1}mM^{-1})	365 ± 44.5	8.3E-2	
Phenylglyoxal			
k_{cat} (min^{-1})		2.4	64.7
K_m (mM)		35 ± 0.1	7.9 ± 5
k_{cat}/K_m (min^{-1}mM^{-1})		6.9E-2	8.2
Acrolein			
k_{cat} (min^{-1})	37.6	1.0	NMA
K_m (mM)	0.802 ± 0.21	22 ± 5.2	
k_{cat}/K_m (min^{-1}mM^{-1})	47.0	4.6E-2	
Butyraldehyde			
k_{cat} (min^{-1})	25.4	0.5	NMA
K_m (mM)	0.06 ± 0.005	17.5 ± 8	
k_{cat}/K_m (min^{-1}mM^{-1})	439	2.8E-2	
NADPH			
k_{cat} (min^{-1})	70.1 ± 1.2	123	37.3
K_m (mM)	0.002 ± 0.0002	0.37 ± 0.1	0.007 ± 0.001
k_{cat}/K_m (min^{-1}mM^{-1})	30900 ± 2600	339	5328

NOTE: Activity measurements with purified recombinant yeast aldo-keto reductases were carried out in a Tris-MES buffer at pH 8.0. Kinetic constants with recombinant human aldose reductase were reported elsewhere (17,22,23). NMA, no measurable activity; Trace indicates that detectable activity was below the level necessary to accurately determine kinetic constants.

rates were observed using phenylglyoxal (apparent $K_m \sim$ 8 mM). Using this substrate, we were able to determine the apparent $K_{m\ NADPH}$ to be approximately 7 μM. Substitution of NADPH with NADH as the pyridine cofactor failed to support reductase activity. Further study will be required to identify additional compounds that could be physiologically significant substrates for YDL124Wp.

Comparative DNA sequence analysis of YJR096W and YDL124W in *Saccharomyces* species

Johnston and coworkers have recently established the genome sequence of several *Saccharomyces* species in an effort to identify functional elements by comparative DNA sequence analysis (16). Since the functions of YJR096W and YDL124W are not currently recognized, we hypothesized that an analysis of the encoded protein sequences deduced from various *Saccharomyces* species would suggest whether the respective protein products were of functional significance. As shown in Table 3, the yeast gene products that display robust aldo-keto reductase activity (GRE3p, GCY1p, YPR1p, YJR096Wp) are all highly conserved across *Saccharomyces* species (88-98% identity to their *Saccharomyces cerevisiae* orthologs). Although YDL124W displays only modest reductase activity *in vitro,* a high degree of conservation is seen with the gene product in *S. paradoxus* (92%) and *S. mikatae* (89% identity). The genomic region corresponding to the ortholog in *S. bayanus* requires further coverage before a homology calculation can be completed.

Catalytically competent mammalian AKRs all possess a catalytic tetrad consisting of aspartate[43], tyrosine[48], lysine[77] and histidine[110] (numbering according to the human aldose reductase primary sequence), with tyrosine[48] and histidine[110] playing key roles in the reaction mechanism of aldehyde reduction (17,18). As shown in figures 1 and 2, alignment of the human aldose reductase primary sequence with YJR096Wp and YDL124Wp derived from *Saccharomyces* species demonstrates that all four catalytic residues have been completely conserved. Indeed, all four catalytic residues are conserved in each of the six putative AKRs identified in *Saccharomyces cerevisiae* (2).

Phenotypic Characterization of Aldo-Keto Reductase Knock Out Strains

In an effort to determine the physiological function of the AKRs in *Saccharomyces cerevisiae*, both individually and as a family of enzymes, we constructed isogenic deletion strains lacking one or more AKR open reading frames. As noted in our previous studies, AKR null strains containing a single

Table 3. Homology of orthologous aldo–keto reductases in *Saccharomyces* species.

	GRE3	GCY1	YPR1	YJR096W	YDL124W
S. paradoxus	98	94	95	95	92
S. mikatae	96	89	92	91	89
S. bayanus	93	88	93	91	-

NOTE: Data are percent identity with the corresponding ortholog in *Saccharomyces cerevisiae*; "-" Data for the YDL124W ortholog in *S. bayanus* are not currently available. Genome sequences that made these comparisons possible were generously provided by P. Cliften, M. Johnston and the Washington University Genome Sequencing Center (personal communication).

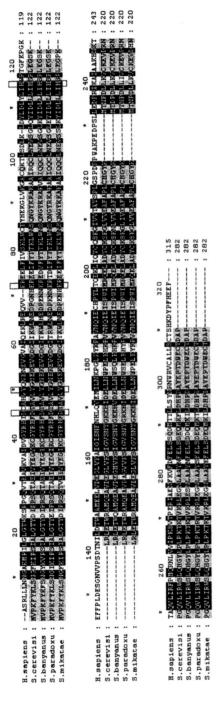

Figure 1. Alignment of *Saccharomyces* orthologs of YJR096W.

Sequences were aligned using Vector NTI (Informax) and GeneDoc. Aligned residues demonstrating 100% conservation with human aldose reductase are shown in black background; those demonstrating 80-99% conservation are shown in gray background. Key catalytic residues identified in mammalian AKRs are shown in boxes.

234

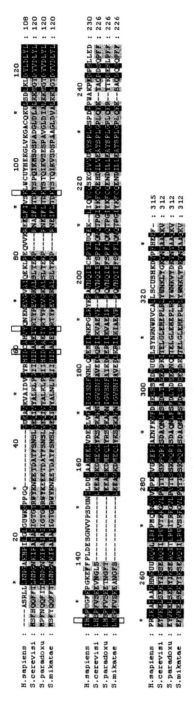

Figure 2. Alignment of *Saccharomyces* orthologs of YDL124W

Sequences were aligned using Vector NTI (Informax) and GeneDoc. Aligned residues demonstrating 100% conservation with human aldose reductase are shown in black background; those demonstrating 80-99% conservation are shown in gray background. Key catalytic residues identified in mammalian AKRs are shown in boxes.

AKR open reading frame deletion are without noticeable phenotype when screened for sensitivity to a variety of stress conditions such as osmotic and oxidative stress and heat shock (2,6). We hypothesized that yeast strains null for a single AKR open reading frame were phenotypically normal due to functional compensation by remaining AKR genes. Support for this hypothesis can be drawn from the broad overlapping substrate specificity observed with at least three of the six yeast AKRs (GRE3p, YPR1p, and GCY1p) (2). Accordingly, we devised a scheme whereby open reading frames corresponding to these three catalytically active AKRs were progressively deleted.

As shown previously, isogenic double AKR null strains (*gre3Δ ypr1Δ*, *gre3Δ gcy1Δ, ypr1Δgcy1Δ*) were all found to be without significant phenotype (6). However, the triple AKR null strain *gre3Δ ypr1Δ* gcy1Δ demonstrated a marked sensitivity to heat shock (6). To establish the effect of AKR gene deletion on yeast metabolism and stress response, we compared the transcriptional profile of wild type and the isogenic triple AKR null strain using Affymetrix oligonucleotide arrays.

DNA microarray studies

Given the marked sensitivity of the triple AKR null strain (*gre3Δ ypr1Δ* gcy1Δ) to heat shock stress, we compared the transcription profile of wild type and triple null strains in log phase growth under both normal conditions (30°C) and following a mild heat shock (30 min, 37°C). Each strain was cultured in triplicate under both growth conditions (12 microarray chips in total). DNA targets were produced from total RNA extracted from yeast cultures using standard conditions.

Data from the microarray chips (Affymetrix Genechip YGS98) were initially subjected to a scaling process using Affymetrix Microarray Suite 5.0 software. The resulting data were then imported into Silicon Genetics GeneSpring 4.2. Each ORF was normalized to a value of 1 using the mean of the values obtained from all 12 gene chips. As only gene sets that passed a comparatively more stringent significance test were entered into the results reported here, fewer induced gene transcripts are identified than were previously reported from this study (6).

A comparison of the transcription profile from wild type *Saccharomyces cerevisiae* grown under normal and heat shock conditions shows that heat shock results in the induction of a large number of genes. As shown in Figure 3, 346 genes showed a ≥2.5-fold increase in transcript level. In contrast, only 61 genes met the same criteria for induction following heat shock in the triple AKR null strain. Of the 61 heat shock-induced genes, the vast majority could be classified as heat shock responsive, as they were the same genes as those induced by heat shock in the wild type strain. Nevertheless, it is clear that the triple AKR null strain has a markedly reduced ability to mount a transcriptional response to heat shock stress.

Comparison of wild type and the triple AKR null strains grown under basal culture conditions also revealed an unexpected transcriptional profile. A large number (91) of genes appear to be activated even under basal culture conditions in the triple null strain as compared to wild type (Figure 4). All 91 genes found to be constitutively-induced in the triple AKR null strain were heat shock inducible in the wild type strain. However, of the 91 constitutively-induced genes in the triple AKR null strain, only 17 were further inducible by heat shock in the deletion strain (data not shown).

These data indicate that ablation of the three most catalytically competent aldo-keto reductase genes results in dysregulation of the heat shock response in *Saccharomyces cerevisiae*. Many genes that should be induced by heat shock are constitutively upregulated even under basal conditions. Furthermore, heat shock treatment of the triple AKR null strain fails to elicit the large majority of genes that should respond to this type of stress, as evidenced by the relatively small number of genes that are induced (~18% of wild type). This combination of dysregulated transcription control most likely is responsible for the heat shock sensitivity we observed in the triple AKR null strain.

Many investigators have shown that the stress response in yeast is mediated in part through the action of the transcription activators Msn2p and Msn4p (15,19). The ability of these transcription activators to affect gene expression is controlled in part through their subcellular distribution. To upregulate target genes, Msn2p and Msn4p, which are cytoplasmic proteins, must accumulate in the nucleus. Nuclear localization is reported to be affected by protein kinase A-mediated phosphorylation (5) or alternatively by interactions with factors that retain them in the cytoplasm (20). It is interesting to consider whether the altered transcription profile observed in the triple AKR null strain results from an alteration in the subcellular distribution of transcriptional activators such as Msn2p and Msn4p or other factors that contribute to the regulation of the stress response. Additional studies will be required to establish a mechanism linking the presence of the yeast AKRs and the stress response program in yeast.

References

1. The Diabetes Control and Complications Trial Research Group. The Diabetes Control and Complications Trial. *Arch. Ophthal.* **1995**, *113*, 36-51.
2. Petrash, J. M.; Murthy, B. S.; Young, M.; Morris, K.; Rikimaru, L.; Griest, T. A.; Harter, T. *Chem.Biol.Interact.* **2001**, *130-132*, 673-683.

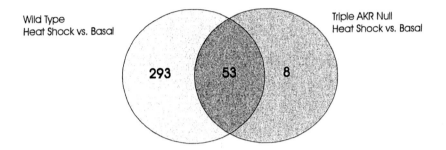

Figure 3. The triple AKR null strain has an attenuated heat shock response.

Heat shock resulted in induction of 346 genes in the wild type strain compared with 61 genes in the triple AKR null strain. The region of overlap shows that 53 genes in common were induced in both strains.

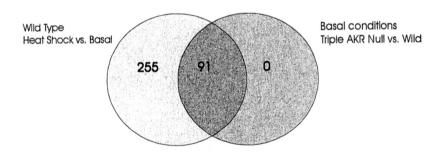

Figure 4. Constitutive upregulation of heat shock responsive genes in the triple AKR null strain.

Heat shock resulted in induction of 346 genes in the wild type strain. When compared with wild type under basal culture conditions, the triple AKR null strain demonstrated upregulation of 91 genes. The region of overlap shows that all 91 upregulated genes in the null strain were heat shock inducible genes in the wild type.

3. Kuhn, A.; van Zyl, C.; van Tonder, A.; Prior, B. A. *Appl.Environ. Microbiol.* **1995**, *61*, 1580-1585.
4. Garay-Arroyo, A.; Covarrubias, A. A. *Yeast* **1999**, *15*, 879-892.
5. Garreau, H.; Hasan, R. N.; Renault, G.; Estruch, F.; Boy-Marcotte, E.; Jacquet, M. *Microbiology* **2000**, *146 (Pt 9)*, 2113-2120.
6. Chang, Q.; Harter, T. M.; Rikimaru, L. T.; Petrash, J. M. Chem. Biol. Interact. 2002, (*In Press*)
7. Oechsner, U.; Magdolen, V.; Bandlow, W. *FEBS Letts.* **1988**, *238*, 123-128.
8. Carper, D.; Nishimura, C.; Shinohara, T.; Dietzchold, B.; Wistow, G.; Craft, C.; Kador, P.; Kinoshita, J. H. *FEBS Letts.* **1987**, *220*, 209-213.
9. Hur, E.; Wilson, D. K. *Acta Crystallogr. D. Biol. Crystallogr.* **2000**, *56 (Pt 6)*, 763-765.
10. Hur, E.; Wilson, D. K. *Chem. Biol. Interact.* **2001**, *130-132*, 527-536.
11. Nakamura, K.; Kondo, S.; Kawai, Y.; Nakajima, N.; Ohno, A. *Biosci. Biotechnol. Biochem.* **1997**, *61*, 375-377.
12. Ford, G.; Ellis, E. M. *Yeast* **2002**, *19*, 1087-1096.
13. Norbeck, J.; Blomberg, A. *J. Biol. Chem.* **1997**, *272*, 5544-5554.
14. Kim, S. T.; Huh, W. K.; Lee, B. H.; Kang, S. O. *Biochim. Biophys. Acta* **1998**, *1429*, 29-39.
15. Causton, H. C.; Ren, B.; Koh, S. S.; Harbison, C. T.; Kanin, E.; Jennings, E. G.; Lee, T. I.; True, H. L.; Lander, E. S.; Young, R. A. *Mol. Biol. Cell* **2001**, *12*, 323-337.
16. Cliften, P. F.; Hillier, L. W.; Fulton, L.; Graves, T.; Miner, T.; Gish, W. R.; Waterston, R. H.; Johnston, M. *Genome Res.* **2001**, *11*, 1175-1186.
17. Tarle, I.; Borhani, D. W.; Wilson, D. K.; Quiocho, F. A.; Petrash, J. M. *J. Biol. Chem.* **1993**, *268*, 25687-25693.
18. Bohren, K. M.; Grimshaw, C. E.; Lai, C.-J.; Harrison, D. H.; Ringe, D.; Petsko, G. A.; Gabbay, K. H. *Biochemistry* **1994**, *33*, 2021-2032.
19. Estruch, F. *FEMS Microbiol. Rev.* **2000**, *24*, 469-486.
20. Beck, T.; Hall, M. N. *Nature* **1999**, *402*, 689-692.
21. Hodges, P. E.; Payne, W. E.; Garrels, J. I. *Nucleic Acids Res.* **1998**, *26*, 68-72.
22. Petrash, J. M.; Harter, T. M.; Devine, C. S.; Olins, P. O.; Bhatnagar, A.; Liu, S. Q.; Srivastava, S. K. *J. Biol. Chem.* **1992**, *267*, 24833-24840.
23. Srivastava, S.; Watowich, S. J.; Petrash, J. M.; Srivastava, S. K.; Bhatnagar, A. *Biochemistry* **1999**, *38*, 42-54.

Indexes

Author Index

Subject Index

248

M

N

RETURN TO: **CHEMISTRY LIBRARY**
100 Hildebrand Hall · 510-642-3753

LOAN PERIOD 1	2	3
1-MONTH USE		
4	5	6

ALL BOOKS MAY BE RECALLED AFTER 7 DAYS.
Renewals may be requested by phone or, using GLADIS,
type **inv** followed by your patron ID number.

DUE AS STAMPED BELOW.

NON-CIRCULATING
UNTIL: _____ MAY 04 2004

FORM NO. DD 10
2M 4-03

UNIVERSITY OF CALIFORNIA, BERKELEY
Berkeley, California 94720–6000